JPL Publication 02-5

I0393899

Radio Wave Propagation Handbook for Communication on and Around Mars

Christian Ho
Nasser Golshan
Arvydas Kliore

National Aeronautics and
Space Administration

Jet Propulsion Laboratory
California Institute of Technology
Pasadena, California

March 1, 2002

Abstract

This handbook examines the effects of the Martian environment on radio wave propagation on Mars and in the space near the planet. The environmental effects include these from the Martian atmosphere, ionosphere, global dust storms, aerosols, clouds, and geomorphologic features. Relevant Martian environmental parameters were extracted from the measurements of Mars missions during the past 30 years, especially from Mars Pathfinder and Mars Global Surveyor. The results derived from measurements and analyses have been reviewed through an extensive literature search. The updated parameters have been theoretically analyzed to study their effects on radio propagation. This handbook also provides basic information about the entire telecommunications environment on and around Mars for propagation researchers, system engineers, and link analysts. Based on these original analyses, some important recommendations have been made, including the use of the Martian ionosphere as a reflector for Mars global or trans-horizon communication between future Martian colonies, reducing dust storm scattering effects, etc. These results have extended our wave propagation knowledge to a planet other than Earth. The tables, models, and graphics included in this handbook will benefit telecommunication system engineers and scientific researchers.

Acknowledgements

This work was performed at the Jet Propulsion Laboratory, California Institute of Technology, Pasadena, under contract with the National Aeronautics and Space Administration. This work was sponsored by NASA's Cross Cutting Technology Development Program and the Deep Space Communications and Navigation Systems at Jet Propulsion Laboratory.

The Authors would like to thank Steve Slobin, Miles Sue, Anil Kantak, and Eni Njoku of JPL for their support and advice in the preparation of this handbook. We also thank Roger Carlson, technical editor of JPL's Technology Information Section, for coordinating the publication of this handbook.

Table of Contents

1. **Introduction** 1
   1.1   The Mars Environment 1
   1.2   Radio Wave Propagation Parameters.. 4

2. **Martian Ionosphere and Its Effects on Propagation (Plasma and Magnetic Field)**.. 7
   2.1   Introduction.. 7
   2.2   Formation of the Martian Ionosphere 8
   2.3   Dayside Martian Ionospheric Structure 9
   2.4   The Nightside Martian Ionosphere 13
   2.5   Ionospheric Effects on Radio Wave Propagation 14
   2.6   Summary and Recommendations 18

3. **Martian Atmosphere and Its Effects on Propagation** 21
   3.1   Introduction.. 21
   3.2   Martian Tropospheric Effects 23
   3.3   Martian Clouds and Fogs 30
   3.4   Martian Aerosols... 34
   3.5   Communication Blackout During Atmospheric Entry Phase 36
   3.6   Summary and Recommendations 41

4. **Martian Atmospheric Gaseous Attenuation**... 45
   4.1   Introduction.. 45
   4.2   Martian Gaseous Composition and Comparison With Earth Atmosphere 47
   4.3   Martian Atmospheric Absorption Effects on Microwaves 53
   4.4   Summary and Recommendations 55

5. **Martian Dust Storms and Their Effects on Propagation** 59
   5.1   Introduction.. 59
   5.2   Local and Regional Dust Storms 62
   5.3   Global Dust Storms... 64
   5.4   Effects on Radio Wave Propagation.. 65
        5.4.1 Dust Storm Parameters 65
        5.4.2 Radio Wave Attenuation through Dust Storms 67
   5.5   Summary and Recommendations 69

6. **Martian Geomorphologic Effects on Propagation**.................................... 73
   6.1   Introduction.. 73
   6.2   Mars Polar Ice Caps.. 77
   6.3   Mars "Grand Canyon" Valles Marineris.................................... 82
   6.4   Summary and Recommendations 83

7. **Propagation Issues for Communication Between Earth and Mars**............. 89
   7.1   Free Space Loss Between Mars and Earth.................................. 89
   7.2   Combined Propagation Losses Under Normal and Worst Conditions 89

8.   Summary and Conclusions **95**
    8.1   Ionospheric Effects 95
    8.2   Tropospheric Effects.. 96
    8.3   Gaseous Attenuation 97
    8.4   Dust Storm Effects 97
    8.5   Surface Geomorphologic Structures.. 97
    8.6   Links between Mars and Earth 98
    8.7   Recommendation for Telecommunication Systems Engineer 98
          8.7.1   Martian Ionospheric Effects.. 98
          8.7.2   Martian Atmospheric Effects... 99
          8.7.3   Martian Cloud Effects on Wave Propagation 99
          8.7.4   Martian Atmospheric Gaseous Attenuation............................. 100
          8.7.5   Martian Dust Storm Effects 101
          8.7.6   Communication Blackout during the Martian
                   Atmospheric Entry Phase.. 101

9.   Acronyms **103**

10.   Index... **105**

List of Figures

1-1   Mars Orbit and Relative Distance from Sun 2
2-1   A Martian Ionospheric Altitude Profile of Electron Density Measured
        by Viking Lander 2 7
2-2   Illustration of the Steps that Lead to the Formation of the Mars Ionosphere 9
2-3   Electron and Magnetic Field Observations for MGS Day 262 (Orbit 5) 11
2-4   Peak Electron Densities and Peak Altitudes of the Mars Ionosphere 11
2-5   Martian Ionospheric Plasma Density and Temperature Profiles 12
2-6   Locations of the Top of the Martian Ionosphere from Radio-Occultation Profiles 13
2-7   Mars Nightside Ionosphere Electron Density Profiles at Different
        Solar-Zenith Angles 13
2-8   The Calculated Martian Dayside Ionospheric Altitude Profiles for Different
        Solar-Zenith Angles Using Equation 2-1 15
2-9   The Dayside Martian Ionosphere as a Reflector for Trans-horizon
        Surface-to-Surface Communication 16
3-1   Nominal Northern Summer Midlatitude Model of the Atmosphere of Mars 23
3-2   Atmospheric Pressure Profile Measured by MGS Radio Occultation 24
3-3   The Atmospheric Density Profiles Derived from the Mars Pathfinder Accelerometer
        Data.. 25
3-4   Martian Atmospheric Temperature Profile Measured by MGS Radio Occultation 25
3-5   Models of Martian Atmospheric Surface Temperature Variation and Temperature
        Profiles in the Lowest 8 km 26
3-6   Seasonal Variation of Surface Pressure at the Two Viking Sites 27
3-7   Radio Refractivity for Martian Atmosphere 29

3-8   Temperature Oscillations Found in the Viking Temperature Soundings 29

3-9   A Hubble Telescope Image of Martian Clouds 31

3-10   Simplified Schematic Drawing of the Dust Devil That Passed Over the
       Sagan Memorial Station (Mars Pathfinder) on Sol 25 36

3-11   Diagrammed View of a Blunt Hypersonic Spacecraft Entering the
       Martian Atmosphere. 38

3-12   Electron Densities in the Capsule Wake Region versus Time from Entry for
       Various Entry Velocities and for Entry Angle $\psi = 90°$ 39

3-13   Tracking Signals During Mars Pathfinder Atmospheric Entry Phase 40

4-1   Specific Gaseous Attenuation for a One-Way Horizontal Earth's Atmospheric Path
       in the Frequency Ranges from Microwave to Visible Light...................................... 46

4-2   Martian Atmospheric Density Profiles for Various Constituents 48

4-3   Earth Atmospheric Number Density Profiles for Individual Species 49

4-4   Distribution of Water Vapor in the Martian Atmosphere by Latitude and Season 52

4-5   Gaseous Specific Absorption Attenuation by Water Vapor, Oxygen, and Both
       at the Surface of Earth and Mars 54

5-1   Threshold Velocities for Initiation of Particle Movement as a Function of
       Particle Diameter 59

5-2   A Local Dust Storm Observed by Mariner 9 at the Edge of the South Polar Ice Cap 61

5-3   A Local Dust Storm in the Solis Planum Region at $L_s = 227°$ 61

5-4   A Dust Storm Picture Taken by MGS Orbiter Camera (MOC) During Orbit 235 63

5-5   A Regional Dust Storm Observed by MGS Orbiter Camera (MOC) from Orbit 50 63

5-6   Expansion of the June 1977 Storm, as Derived from Viking Orbiter Visual Imaging 65

5-7   Visible Optical Depths Derived from Viking Lander 1 Measurements 66

6-1   Two Lambert Maps Show the Martian Surface Features 75

6-2   An Image Mosaic Showing the Mars Pathfinder Landing Area 76

6-3   An Image Taken by MGS Showing Signs of Water Erosion and Debris Flow 76

6-4   Mosaics of Viking Orbiter Images of Polar Regions in the Southern Summer 78

6-5   Three-Dimensional View of the North Polar Region of Mars from MGS Orbiter 79

6-6   Graphic Diagram Showing Radio Ray Paths Between a Rover Within a Canyon
       and a Satellite Orbiting Mars 80

6-7   A Color Image of Valles Marineris, the Great Canyon of Mars 84

6-8   Opacity of the Martian Atmosphere Over Valles Marineris Canyon from MOLA 85

6-9   Artist's Conception of a Mars Airplane Flying over Valles Marineris 85

6-10   Geometric Plot Showing the Ray Paths Between a Satellite and a Mars Airplane 86

7-1   Telecommunication Links Around Mars from the Point of View of Radio Wave
       Propagation 92

8-1   Martian Ionospheric Effects on One-Way Radio Wave Propagation of Various
       Frequencies 99

8-2   Martian Gaseous Attenuation for a One-Way Radio Wave Path through the
       Atmosphere for Two Different Elevation Angles (30° and 90°) 100

8-3   Martian Dust Attenuation for One-Way Radio Wave Path through a Dust Cloud
       for Various Elevation Angles and Dust Particle Sizes 101

List of Tables

1-1   Mars Statistical Parameters 1

1-2   Mars Exploring Missions and Active Dates 3

2-1   Ionospheric Peak Density and Critical Frequency for Mars and Earth 15

2-2   Usable Critical Frequency and Hop Distance for Various Launch Angles 16

2-3   Effects of the Total Vertical Electron Content of the Mars Ionosphere on Wave Propagation Characters 17

3-1   Nominal Summer-Seasonal Midlatitude Martian Atmospheric Model 22

3-2   Values for Complex Relative Dielectric Permittivity of Water Ice 32

3-3   One-way Attenuation Coefficient K_i in Clouds 32

3-4   Visual Optical Depths of Clouds and Fogs on Earth and Mars 34

3-5   Critical Plasma Densities and Communication Frequencies 37

4-1   Surface Atmospheric Parameters at Mars and Earth 50

4-2   A Comparison of Atmospheric Compositions Near the Surfaces of Mars and Earth 51

4-3   Ratios of Atmospheric Constituents between Earth and Mars 53

5-1   Martian Great Dust Storms 64

5-2   Dielectric Permittivity Index of Dust Particles 67

5-3   A Comparison of Dust Storm Parameters Between Earth and Mars 69

7-1   Free Space Losses for Various Frequencies Between Mars and Earth 89

7-2   Radio Wave Attenuation Around Mars for Various Frequency Bands 91

7-3   Attenuation for All Possible Links Between Mars and Earth 92

1. Introduction

1.1 The Mars Environment

Mars, the fourth planet from the Sun, is one of the terrestrial planets and Earth's outer neighbor. Throughout history, stargazers often referred to Mars as the Red Planet because its bright appearance and reddish color stand out in the night sky. The rocks, soil, and sky have a red or pink hue. Impressive surface features, such as enormous volcanoes and valleys, are frequently obscured by huge dust storms. Mars is only about half the diameter of Earth. Although the length of the Martian day is almost the same as Earth's, its mass is only 11% of Earth's. Some of the basic Mars statistical data are listed in Table 1-1.

Table 1-1. Mars Statistical Parameters

Parameter	Value
Diameter	6,785 km (4,217 miles)
Length of Day	24 hr 37 min
Mass	0.11 x Earth
Length of Year	687 Earth days
Density (water=1)	3.9
Tilt of Axis	25° 12"
Minimum Distance from Sun	205 million km (128 million miles)
Maximum Distance from Sun	249 million km (155 million miles)
Surface Gravity	0.38 x Earth
Temperature	–82°C to 0°C (–116°F to 32°F)
Minimum Distance from Earth (opposition)	55 million km
Maximum Distance from Earth (superior conjunction)	~400 million km
Satellites	Deimos (8 km) Phobos (28x20 km)

Before space exploration, Mars was considered the best candidate for harboring extraterrestrial life because the Martian environment was expected to be the most similar to Earth. In July of 1965, Mariner 4 transmitted 22 close-up pictures of Mars. All these pictures showed a surface containing many craters and naturally occurring channels. In July and September 1976, Viking Landers 1 and 2 touched down on the surface of Mars. The three biology experiments aboard the landers discovered unexpected and enigmatic chemical activity in the Martian soil, but they provided no clear evidence for the presence of living microorganisms in the soil near the landing sites. However, recent examination of the 12 Martian meteorites collected from Antarctica found carbonate minerals. Fossilized organic materials associated with these carbonates provided evidence for possible past life on Mars.

Recently, scientists using data from Mars Global Surveyor (MGS) have observed features that suggest there may be current sources of liquid water at or near the surface of Mars [Malin and Edgett, 2000]. Many images show gullies formed by flowing water and the deposits of soil and rocks transported by these flows. The evidence makes it much more likely that life could exist or could have existed on the planet. Future exploration of Mars will not only help us answer questions such as whether life exists on Mars, but it will also help reveal the origin and evolution of the Solar System [Carr, 1981; Kieffer et al., 1992; Kliore, 1982; Luhmann et al., 1992; Golombek et al., 1997; Albee et al.,1998].

The Martian climate and surface features are more significantly influenced by the shape of the Martian orbit than Earth because Mars has a more elliptical orbit. The Martian orbit eccentricity is 0.093, in contrast to the near-circular Earth orbit (0.017). This high eccentricity affects Mars in a number of ways, as show in Figure 1-1. When Mars is at its perihelion (closest point to the Sun), the southern Martian hemisphere tilts toward the Sun. Thus, the southern hemisphere has a relatively hot and short summer. When Mars is at its aphelion (farthest point from the Sun), the northern Martian hemisphere tilts toward the Sun. Thus, the northern hemisphere has a relatively cold and long summer. A similar effect on Earth causes only a 3-day difference because of the low eccentricity. For this reason, the Martian southern summer peak temperature is ~ 30°C higher than the northern peak temperature. These differences have generated profound effects on Martian atmospheric circulation patterns, surface geomorphologic change, dust storm and polar ice cap formation, etc. Mars also has a higher inclination angle (25° 12") of its rotation axis relative to the normal to its orbital plane (Earth's inclination is 23.5°). However, the effect due to this difference is small.

American and Russian spacecraft have made a steady stream of exploratory flights to Mars with both successes and failures in the past thirty years [Kliore, 1982; Blamont, 1991; Breus, 1992; Golombek et al., 1997; Albee et al., 1998]. Table 1-2 is a subset of the missions that have returned valuable data and of planned future missions.

Recently, the U.S. National Aeronautics and Space Administration (NASA) launched a series of missions in a coordinated program to explore Mars. Both successful missions of Mars Pathfinder (MPF) and Mars Global Surveyor (MGS) greatly enhanced our knowledge and public interest in Mars [Golombek et al., 1997; Albee et al., 1998]. NASA is accelerating the pace of Mars environmental studies and will be sending spacecraft to Mars more frequently.

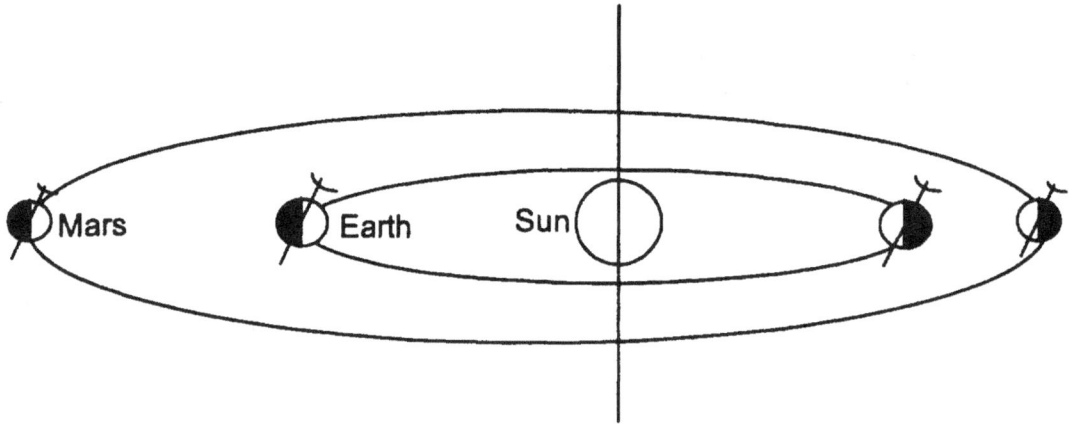

Figure 1-1. Mars Orbit and Relative Distance from Sun. Mars has a more elliptical orbit, as compared with Earth's orbit. There is a 44 million kilometer difference between its aphelion and perihelion.

Table 1-2. Mars Exploring Missions and Active Dates

Missions	Active Date
Mariner 4, 6, 7, and 9 (U.S.)	1964, 1969, 1971
Mars 2 and 3 (Russ.)	1971 & 1972
Mars 4, 5, and 6 (Russ.)	1974
Viking I and II (U.S.)	1975
Phobos (Russ.)	1989
Mars Observer (U.S.)	1993
Mars Pathfinder (U.S.)	1997
Mars Global Surveyor (U.S.)	1998
Mars '98 (Polar Lander and Deep Space II) (U.S.)	1999
Nozomi (Planet-B) (Japan)	1999
Mars Odyssey Orbiter (U.S.)	2001
Mars Exploration Rover (U.S.)	2003
Mars Express Orbiter (ESA) and Beagle II Lander (UK)	2003
Comm Orbiter (U.S.)	2005
Mars Reconnaissance Orbiter (U.S.)	2005
Mars 2005 (Sample Return) (U.S.) and 4 Netlanders (CNES)	2005
Mobile Science Laboratory (U.S. and ESA)	2007
Scout Missions (U.S. and ESA)	2007
Mars Aircraft (Kitty Hawk) (U.S.)	2010
Mars Human Exploration Program (U.S.)	2020

Through cumulative effects, the instruments carried by all Mars missions have provided precise and definitive measurements and analysis. Today our knowledge about Mars has been greatly enhanced in a range from its surface rocks to its atmosphere. The atmosphere of Mars is quite different from that of Earth [Keating et al., 1998]. While the Earth atmosphere is composed mostly of nitrogen and oxygen, the Martian atmosphere is composed primarily of carbon dioxide with small amounts of other gases. The six most common components of the Martian atmosphere are the following:

(1) carbon dioxide (CO_2): 95.32%;
(2) nitrogen (N_2): 2.7%;
(3) argon (Ar): 1.6%;
(4) oxygen (O_2): 0.13%;
(5) water (H_2O): 0.03% and
(6) neon (Ne): 0.00025%.

The Martian atmosphere contains only about 1/1,000 as much water as Earth's, but even this small amount can condense out, forming clouds that ride high in the atmosphere or swirl around the slopes of towering volcanoes. Local patches of early morning fog can form in valleys. At the

Viking Lander 2 site, a thin layer of water frost covered the ground each winter. Local and global dust storms, which occur frequently in certain areas, contribute to the atmospheric hazes [Martin, 1984].

The Martian sky usually appears as a reddish color. This is because the blue light is absorbed by the dust, but the red light is scattered throughout the sky. By contrast, the molecules in the Earth's atmosphere intercept about as much of the blue sunlight as the Martian dust does because blue light is scattered easily by Earth's atmosphere, but the red light is neither absorbed and scattered, giving the Earth its blue sky.

On the basis of previous missions and the recent Mars Pathfinder measurements, there is evidence that in the past a denser Martian atmosphere may have allowed water to flow on the surface. Physical features closely resembling shorelines, gorges, riverbeds, and islands suggest that great rivers once marked the planet. The average recorded temperature on Mars is $-63°C$ ($-81°F$) with a maximum temperature of $20°C$ ($68°F$) and a minimum of $-140°C$ ($-220°F$). Carbon dioxide, the major constituent of the atmosphere, freezes out to form an immense polar cap, alternately at each pole. The atmospheric pressures at the Viking Lander 2 site were 7.3 and 10.8 millibars (1 mb = 100 Pa). In comparison, the average atmospheric pressure of the Earth is 1000 millibars.

The Martian ionosphere also has some differences and similarities with respect to the Earth's ionosphere [Acuna et al., 1998]. Basically, the Martian ionosphere is a single layer with relative low plasma density. Because Mars has little or no intrinsic magnetic field, ionized gas can directly interact with the solar wind to form a comet-like ionotail in the nightside.

1.2   Radio Wave Propagation Parameters

Telecommunication with the spacecraft is crucial to secure the success of each mission. Thus, we need to study the effects of Martian environments on radio wave propagation and any potential impairment to communication.

Because the environment of Mars is significantly different from Earth's in many aspects from its surface to its outer ionosphere, the effects of the Martian environment on radio wave propagation may be also different. Our understanding of radio wave propagation needs to be expanded based on data from the various Mars missions.

From the viewpoint of classical radio wave theory, radio wave propagation at Mars is controlled by both its ionospheric (plasma) refractive index and its tropospheric (atmosphere) refractive index. These indices govern the propagation direction, the intensity, and the polarization of radio waves.

For low-frequency waves, the refractive index of a medium containing free electrons, with a superimposed static magnetic field, is given by the Appletion-Hartree formula [Budden, 1961]:

$$n^2 = 1 - \dfrac{X}{1 - iZ - \dfrac{Y_T^2/2}{1-X-iZ} \pm \sqrt{\dfrac{Y_T^4/4}{(1-X-iZ)^2 + Y_L^2}}} \qquad (1\text{-}1)$$

where n is the refractive index, $X = \dfrac{\omega_p^2}{\omega^2} = \dfrac{f_p^2}{f^2} = \dfrac{N_p e^2}{\varepsilon_0 \, m \omega^2}$, and $Y = \dfrac{eB}{m\omega}$, $Y_L = Y\cos\theta_{Bk}$, which is the longitudinal component of Y, while $Y_T = Y\sin\theta_{Bk}$ is the transverse component of Y. $Z = v/\omega$. Furthermore, ω is the radio wave angular frequency, ω_p is the plasma frequency, v is the plasma collision frequency, B is the background magnetic field, and θ_{Bk} is the angle between the magnetic field and the wave propagation direction. Thus, the refractive index is mainly a function of electron density and background magnetic field.

For high frequency waves (> 1 GHz), the radiometeorology has some effects on the wave propagation. These effects mainly occur in the lower portion of the atmosphere, the troposphere. Because the tropospheric radio refractive index is slightly greater than unity, it is convenient to define [Bean and Dutton, 1966]:

$$N = (n-1) \times 10^6 \quad \text{(N unit)} \qquad (1\text{-}2)$$

We usually use N (refractivity) to describe the spatial and temporal variation of the air refractive index. In general, the dry part of the refractivity (N_d) can be expressed as:

$$N_d = Q\frac{P}{T} \qquad (1\text{-}3)$$

where $Q = 0.269\sum_i q_i f_i$ and q_i is the refractivity at standard temperature and pressure (S.T.P.) of the ith constituent gas of the atmosphere (cf. Essen and Froome, [1951]; Newell and Baird, [1965]), and f_i is its fractional abundance (by volume). P is in millibars, and T is in Kelvins. Thus, we have the Martian radio refractivity:

$$N = 130.6\frac{P}{T} + 3.73 \times 10^5 \frac{P_{wv}}{T^2} \qquad (1\text{-}4)$$

Thus, the tropospheric radio refractivity is a function of atmospheric pressure, absolute temperature, and water vapor pressure, P_{wv} (mb). Note that this expression is different from that on Earth because the Martian atmosphere has a different composition than that of Earth.

Besides ionospheric and tropospheric refractive index effects on the wave propagation, the gaseous attenuation of high frequency radio waves is another important factor [Waters, 1976]. Martian dust storms and atmospheric aerosols are the dominant factors in wave scattering. Even though Mars has a very low water vapor content, Martian cloud and morning fogs may also have some impact on radio waves. Some special Martian geological and geomorphologic features, such as polar ice caps, canyons, and crater domes, can also cause wave reflection and diffraction.

In the following chapters, we review all related previous measurements and analysis. We apply the radio wave theory to these environmental parameters and study their effects from outer space to the Martian surface over all related topics.

References

Acuna, M.H., et al., Magnetic field and plasma observations at Mars: Initial results of the Mars Global Surveyor Mission, *Science, 279*, 1676, 1998.

Albee, A.L., et al., Mars Global Surveyor Mission: Overview and Status, *Science, 279*, 1671, 1998.

Blamont, J., Phobos-Mars Mission - Preface, Special Issue, *Planet. Space Sci., 39*, R1, 1991.

Bean, B.R., and E.J. Dutton, *Radio Meteorology*, Dover Publications Inc., New York, USA, 1966.

Breus, T.K., *Venus and Mars: Atmospheres, Ionospheres and Solar Wind Interaction*, Monograph 66, AGU, Washington, DC, 1992.

Budden, K.G., *Radio waves in the ionosphere*, Cambridge University Press, Cambridge, UK, 1961.

Carr, M.H., *The surface of Mars*, Yale University Press, 1981.

Essen, L., and K.D. Froome, the refractive indices and dielectric constants of air and its principal constituents at 2400 Mc/s, *Proc. Phys.* London, 64, 862, 1951.

Golombek, A.P. et al., Overview of the Mars Pathfinder Mission and Assessment of Landing Site Predictions, *Science, 278*, 1743, 1997.

Keating, G.M., et al., The structure of the upper atmosphere of Mars: In situ accelerometer measurements from Mars Global Surveyor, *Science, 279*, 1672, 1998.

Kieffer, H.H., et al., *Mars*, Univ. of Arizona Press, Tucson, 1992.

Kliore, A.J., The Mars reference Atmosphere, Special Issue, in COSPAR, Edited by Kliore, A.J., *Adv. Space Res.*, 2, 1982.

Luhmann, J.G., et al., Evolutionary impact of sputtering of the Martian atmosphere by O+ pickup ions, *Geophys. Res. Lett., 19*, 2151, 1992.

Malin, M.C., and K.S. Edgett, Evidence for recent groundwater seepage and surface runoff on Mars, *Science, 288*, 2330, 2000.

Martin, L.J., Clearing the Martian air: The troubled history of dust storms, *Icarus, 57*, 317, 1984.

Newell, A.C., and R.C. Baird, Absolute determination of refractive indices of gases at 47.7 gigahertz, *J. App. Phys., 36*, 3751, 1965.

Waters, J.W., Absorption and emission by atmospheric gases, *Methods of Experimental Physics*, Vol. 12B, *Radio Telescopes*, 142–176, Ed. M.L. Meeks, Academic Press, New York, NY, USA, 1976.

2. Martian Ionosphere and Its Effects on Propagation (Plasma and Magnetic Field)

2.1 Introduction

As an ionized medium, the ionosphere plays a special role in radio wave propagation. The Martian ionosphere differs from Earth's in a number of ways. Due to the greater distance from the Sun at Mars than Earth, the weaker solar radiation flux generates a lower plasma density in the Martian ionosphere. While Earth's ionosphere has four layers, the Martian ionosphere is a single layer of ionized gas that extends from about 100 kilometers to several hundred kilometers above the surface, as shown in Figure 2-1 from Viking Lander 2 direct measurements [Hanson et al., 1977]. Earth's ionosphere is shielded from the solar wind by a strong planetary magnetic field. In contrast, the Mars ionosphere is directly exposed to the solar wind because Mars lacks a strong magnetic field. Presence of a magnetic field can influence the plasma motion within the ionosphere and also affect low frequency radio wave propagation [Cloutier et al., 1969; Vaisberg, 1976; Luhmann et al., 1987; Luhmann et al., 1992].

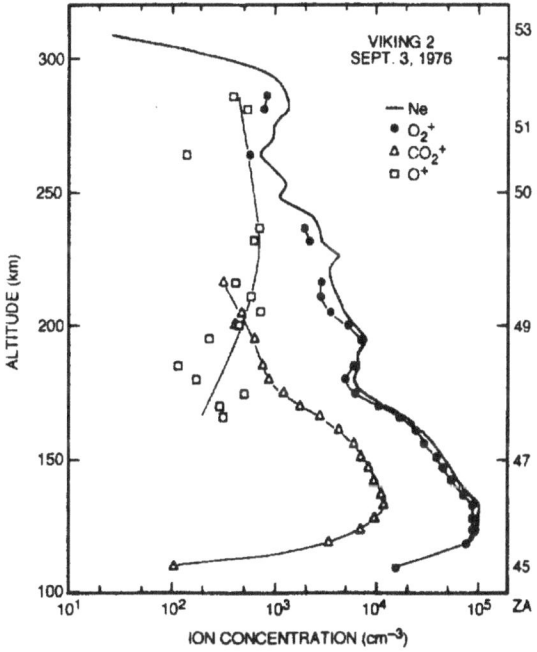

Figure 2-1. A Martian Ionospheric Altitude Profile of Electron Density Measured by Viking Lander 2. Solar zenith angles are marked in the left side. Three ion composition concentrations are also shown (from Hanson et al., 1977).

Previous missions made inconclusive measurements of the Martian magnetic field. The weak magnetic field (< 100 nT) measured by the previous missions had been interpreted as the evidence for a Martian magnetic field [cf. Slavin and Holzer, 1982], although Russell et al. [1984] argued that the measurements could be better explained in terms of a draped interplanetary magnetic field (IMF). Recent measurements by the Mars Global Surveyor (MGS) mission have confirmed that there is no intrinsic dipole magnetic field in Mars [Acuna et al.,

1998]. The MGS magnetometer discovered that the Martian magnetic field is very weak compared to that of Earth's magnetic field, only 1/800 the strength. The weak magnetic field is probably generated by a diffused draping IMF. The solar wind rams into the Martian ionosphere and generates complicated magnetic fields. Thus, this region may have a complicated interaction with the Martian magnetosphere [Acuna et al., 1998; Vaisberg, 1976; Slavin et al., 1991; Woo and Kliore, 1991].

2.2   Formation of the Martian Ionosphere

The extreme ultraviolet (EUV) radiation from the Sun creates the dayside ionosphere of Mars by photo-ionization of its upper atmosphere [McElroy et al., 1977; Fox and Dalgarno, 1979]. This can be described by an ideal Chapman layer, which is formed through the solar wind plasma interaction with a typical unmagnetized planet with atmosphere. The interaction process is shown in Figure 2-2, which is based on a combination of observations and theoretical calculations [Luhmann et al., 1987]. Since there is little or no intrinsic magnetic field on Mars, on dayside, the solar wind can directly interact with its ionosphere [Cloutier et al., 1969]. This forms an extensive comet-like induced magnetosphere, which generally stands off the solar wind well above the planet. The solar wind is shocked and diverted around the ionopause where its incident pressure is approximately balanced by the thermal pressure of the ionospheric plasma. The region between the bow shock and the obstacle is called the magnetosheath or planetsheath. The draped IMF piles up in the front of the ionosphere, forming a region called the "magnetic barrier" [Zhang et al., 1991]. The ions are picked up by the solar wind flow, and they in turn slow the solar wind flow. The shielding current carried by the Mars ionosphere can largely exclude IMF and solar wind plasma from altitudes below the ionopause. The ionopause is also a region of transition between the cold planetary plasma and the post-shocked hot solar wind plasma. This interface between the ionospheric plasma and the shocked solar wind surrounds the planet and extends several thousand kilometers downstream where it forms the ionotail in the nightside [Bauer and Hartle, 1973; Vaisberg and Smirnov, 1986]. When the upstream solar wind carries the IMF through the bow shock and magnetosheath to the obstacle, the field line either slips through or around the conducting ionosphere. This diversion and slowing of the flow, possibly enhanced by mass loading of the flow near the obstacle, causes the field lines to drape and form an ionotail. Since the ionosphere is a conductor, the magnetic field convecting with solar wind plasma generates currents in it that keep the field from penetrating through the Martian ionosphere. Eventually, however, the field diffuses into Mars on a time scale that depends on the ionosphere's conductivity.

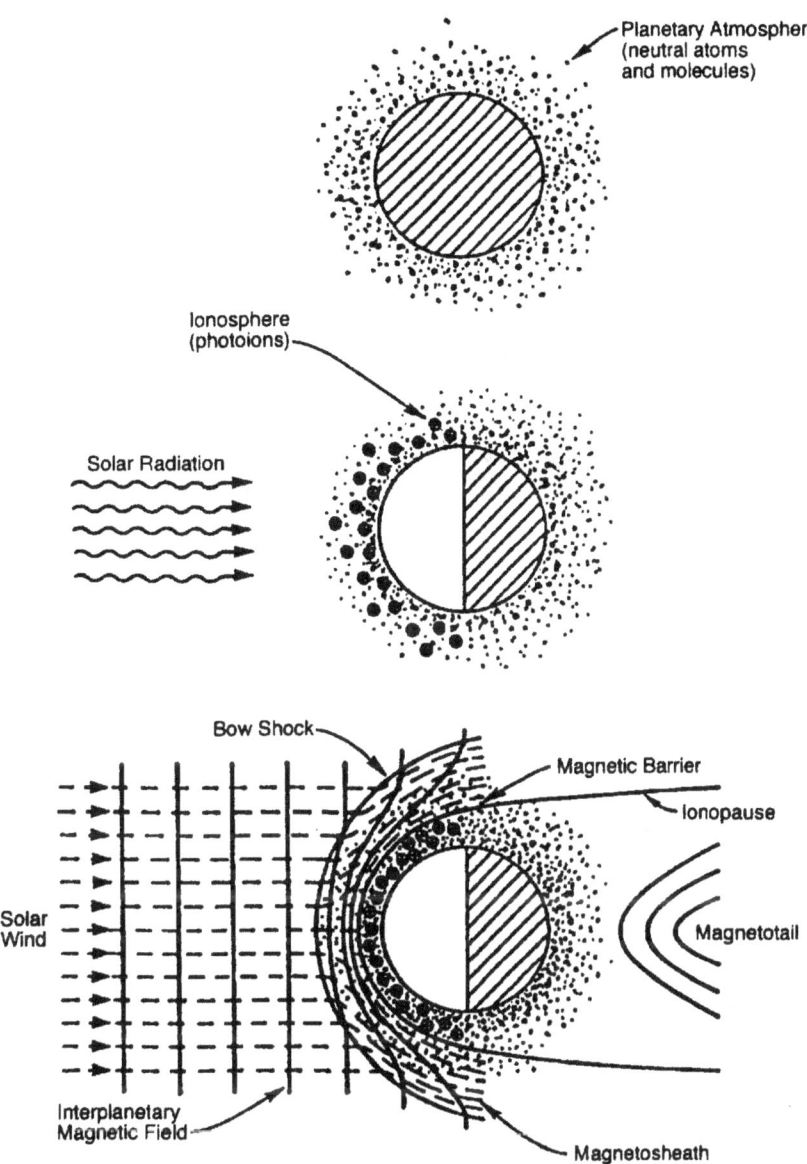

Figure 2-2. Illustration of the Steps that Lead to the Formation of the Mars Ionosphere in the Solar Wind and Interplanetary Magnetic Field (from Luhmann et al., 1987).

2.3 Dayside Martian Ionospheric Structure

Most of the Mars ionospheric measurements were made through radio occultation experiments performed by Mars missions from the United States and the former Soviet Union during the past 30 years [Gringauz, 1976a and b; Hanson and Mantas, 1988; Kliore et al., 1966, 1969, and 1973; Kolosov, 1972]. The only in-situ ionospheric measurements were obtained by two Viking Landers [Hanson et al., 1977; Chen et al., 1978; Mantas and Hanson, 1987] and by MGS [Acuna et al., 1998]. During the Aerobraking phase, the MGS orbiter reached as low as 108 km in altitude, below the peak of the ionosphere. Figure 2-3 shows the measurements made by the MGS Electron Reflector and Magnetometer from its 5th orbit.

When the spacecraft approached Mars, there was a sudden increase in magnetic field associated with the bow shock crossing. Then, in the sheath region, the magnetic field was turbulent and the plasma strongly energized. When the spacecraft was close to the ionosphere, at about 1000 km above the Mars surface, it detected a magnetic pile-up (magnetic barrier) region. Plasma density also increased with decreasing altitude. At about 130 km altitude, the ionosphere reached its density peak. Below the ionosphere, the magnetic field was very low (~5 nT), ruling out the possibility of a planet-wide intrinsic magnetic field. However, occasionally, the MGS magnetometer detected magnetic anomalies as strong as 400 nT near the Martian surface (not shown in Figure 2-3). These small spatial scale anomalies indicated that there are some local magnetic or iron structures in the Mars crust. On the outbound pass, the spacecraft detected similar ionospheric features.

Statistical studies based on available data [Hantsch and Bauer, 1990] indicate that the dayside Martian ionosphere may be generally described using a simple Chapman layer model:

$$N(h) = N_m \exp\left\{0.5\left[1 - (h - h_m)/H - \exp(-(h - h_m)/H)\right]\right\} \qquad (2\text{-}1)$$

where
$$N_m = N_0 (\cos \chi)^k \qquad (2\text{-}2)$$

and
$$h_m = h_0 + H \ln \sec \chi \qquad (2\text{-}3)$$

where N_m is electron peak density, h_m is peak height, χ is solar zenith angle (SZA), H is the scale height of the neutral constituents, and N_0 and h_0 are the peak electron density and the peak height over the zenith ($\chi = 0°$).

For an ideal Chapman layer ($k = 0.5$), the ionosphere should be in photochemical equilibrium with both the neutral gas scale height H and the ionizing radiation flux constant [Ratcliffe, 1972]. The peak heights increase with increasing solar zenith angle toward the terminator. For the real dayside Mars ionosphere, it is found that $k = 0.57$ through a curve fitting [Zhang et al., 1990a]. This small departure from the ideal case is expected because the ionizable constituent and the scale height vary with solar zenith angle and solar activity.

Figure 2-4a shows the data and a best fit curve. The peak electron densities decrease with increasing SZA with a ±15% fluctuation. At 0° SZA, the N_0 is expected to be 2×10^5 cm^{-3}, even though low-zenith-angle ($\chi < 40°$) measurements of the Martian ionospheric electron densities cannot be obtained with radio occultation. Another major factor in determining the ionospheric profile is the density peak height. Figure 2-4b shows the curve fit using $h_m = 120 + 10 \ln \sec\chi$. The peak height is at about 120 km altitude. The Mariner 9 measurements are excluded from this fit because these data significantly departed from the group. High peak altitude values measured in the Mariner 9 mission were caused by the 1971 great Mars global dust storm [Kliore et al., 1973; Stewart and Hanson, 1982]. This global dust storm appears to have elevated the Martian ionosphere as a whole by ~20–30 km without otherwise notably altering its density profile. The heating of the Martian atmosphere by the dust storm enhanced ionospheric scale height and peak height. McElroy et al. [1977] were able to model this behavior by assuming that a 20-K increase in temperature was followed by a steady cooling (8 K/6 weeks).

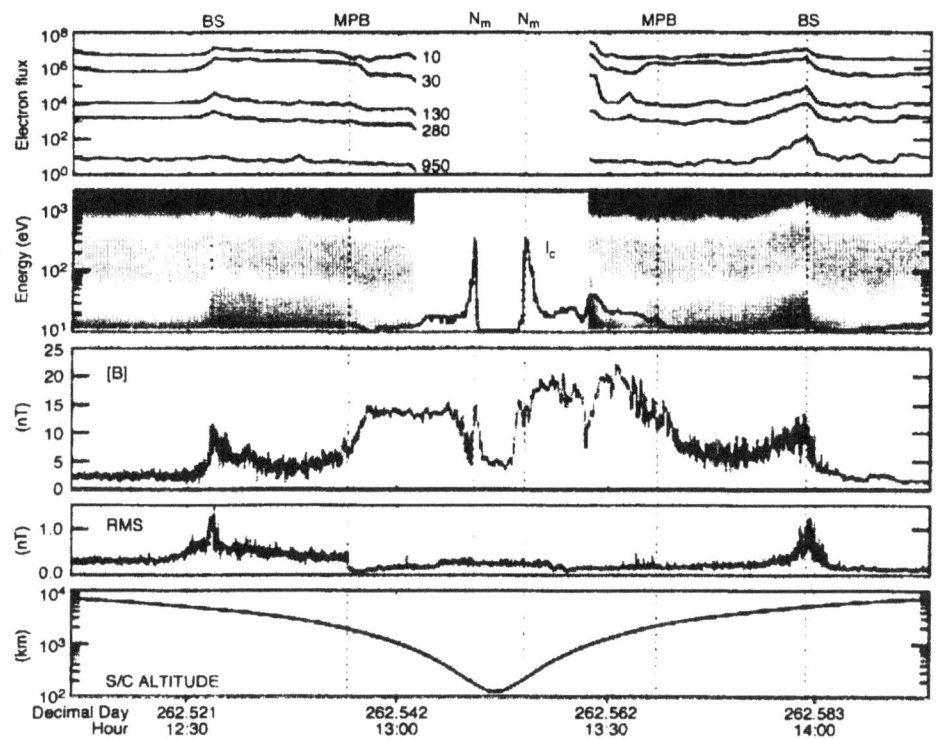

Figure 2-3. Electron and Magnetic Field Observations for MGS Day 262 (Orbit 5).
Electron fluxes are shown as line traces at five energies (upper panel) and as a spectrogram (second panel) with the relative density of cold (<10 eV) electrons superimposed by using the same number of logarithmic intervals as the spectrogram's energy scale. The next three panels show the magnetic field amplitude, and the root mean square (RMS), and the spacecraft altitude. Vertical lines indicate the locations of the bow shock (BS), magnetic pile-up boundary (MPB), and ionospheric main peak (N_m) (from Acuna et al., 1998).

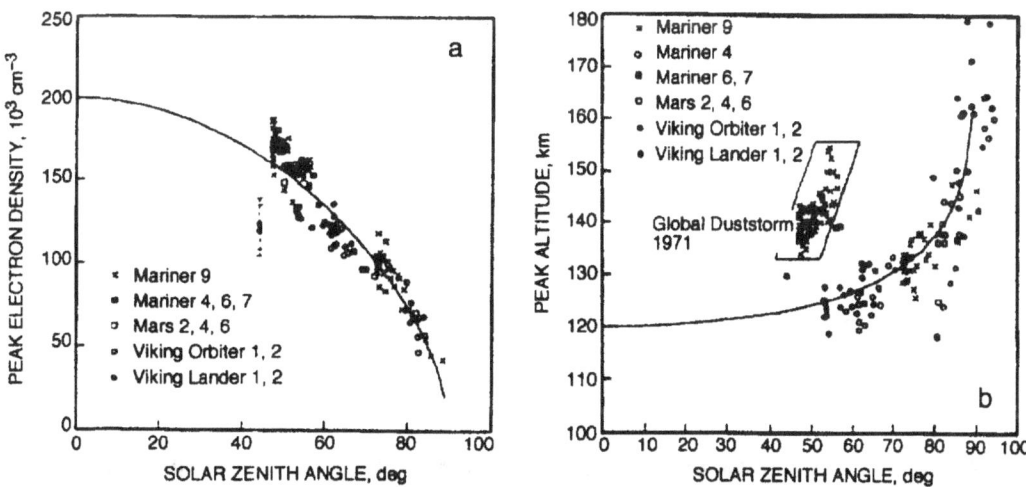

Figure 2-4. Peak Electron Densities and Peak Altitudes of the Mars Ionosphere. (a) Peak electron densities of the Mars ionosphere as a function of solar-zenith angle; (b) Peak altitudes versus solar-zenith angle (from Hantsch and Bauer, 1990).

11

In the lower ionosphere, photoelectron ionization is significant and makes a contribution of 20–30% to the total ionization rate [Nier and McElroy, 1977]. Even though CO_2 is the major atmospheric constituent of Mars at low altitudes and CO_2^+ ions are the primary ions produced below 100 km, O_2^+ ions are dominant at low altitudes because most of the CO_2^+ ions are broken down into O_2^+ ions through a subsequent ion-neutral reaction $(CO_2^+ + O \rightarrow O_2^+ + CO)$. Ion composition profiles and temperature are showed in Figure 2-5a and 2-5b [Shinagawa and Cravens, 1989].

Because previous Mars missions covered more than three solar cycles, the maximum electron density as a function of solar radio flux has been studied from all data [Bauer and Hantsch, 1989; Hantsch and Bauer, 1990]. The peak density can change from 1.0×10^5 cm^{-3} at the solar minimum to 2.5×10^5 cm^{-3} at the maximum [Bauer and Hantsch, 1989]. At solar maximum, the topside ionospheric profile usually has large variations. There is an upper boundary, the ionopause, where solar wind dynamic pressure reaches a balance with the ionospheric thermal pressure and the plasma density falls sharply. When the ionopause is high, there is a fairly large plasma scale height (several hundred kilometers) below the ionopause. The topside ionospheric profile responds sensitively to incident solar wind pressure (and to a lesser extent to changes in the solar EUV flux). In contrast, at the solar minimum, the ionosphere constantly falls off with a small scale height (~20 km), and the ionosphere does not show such a response because the ionopause is within the deep region dominated by photochemistry (below ~225 km). On the basis of the past 30 years of measurements, the dayside Mars ionopause (the upper boundary of the ionosphere) locations detected during various solar cycle are shown in Figure 2-6 [Slavin and Holzer, 1982]. In comparison, the Martian obstacle location (which is inferred from the bow shock shape) is also displayed. The lower limit at solar minimum in 1965 was nearly 250 km from Mariner 4, and the upper limit at solar maximum was almost 400 km. No large solar zenith angle dependence has been seen in the Martian upper ionosphere.

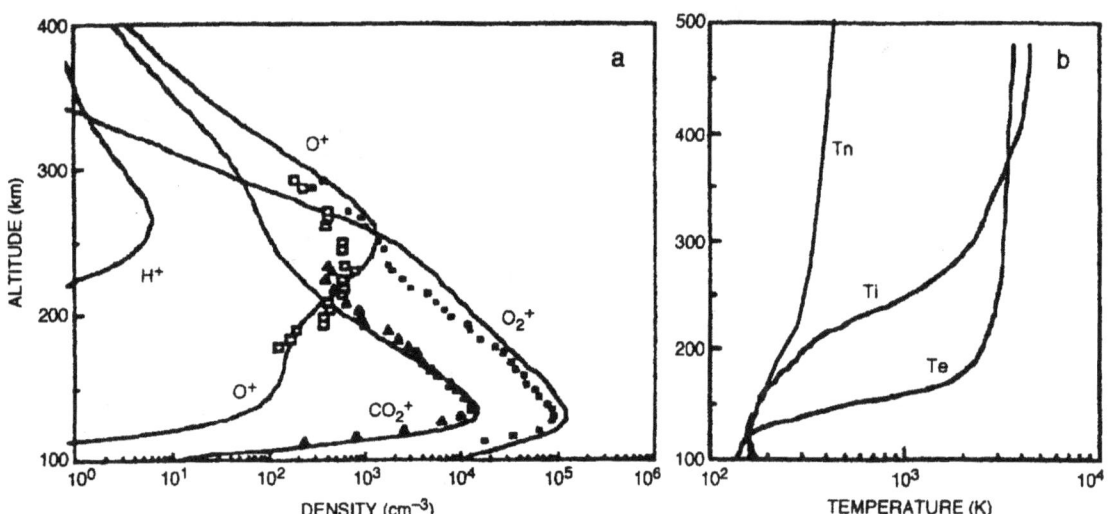

Figure 2-5. Martian Ionospheric Plasma Density and Temperature Profiles. (a) Comparison of the calculated ion density profiles (solid lines from Shinagawa and Cravens, 1989) with Viking 1 and 2 measured profiles (Hanson et al., 1977); (b) Altitude profiles of ion, electron, and neutral temperatures.

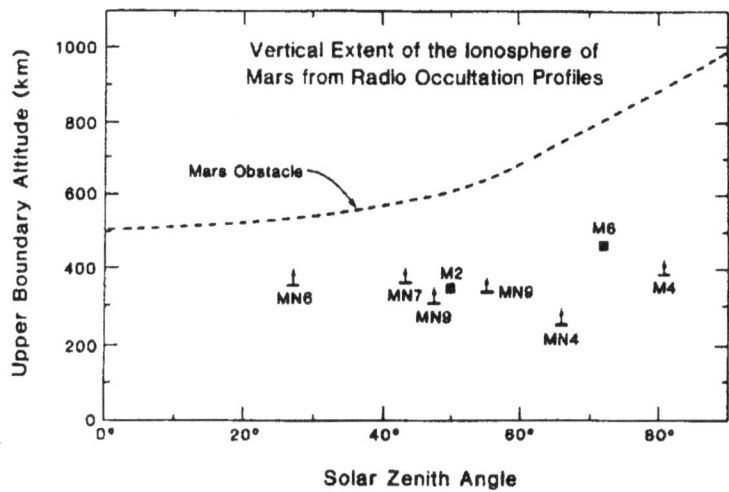

Figure 2-6. Locations of the Top of the Martian Ionosphere from Radio-Occultation Profiles Compared to the Obstacle Height Inferred from Bow-Shock Observations (from Slavin and Holzer, 1982).

2.4   The Nightside Martian Ionosphere

Compared with the dayside ionosphere, there are very few measurements of the Martian nightside ionosphere. Maximum data coverage is from 90° (terminator) to 125° in the nightside SZA [Zhang et al., 1990b]. Figure 2-7 shows three occultation profiles for the nightside ionosphere from Viking 1 orbiter during low solar activity. The plasma density in the nightside ionosphere is frequently too low to be detected by radio occultation. Many profiles from the nightside of Mars do not show any peak at all. The Martian nightside ionosphere usually has an electron density as low as 5×10^3 cm^{-3}.

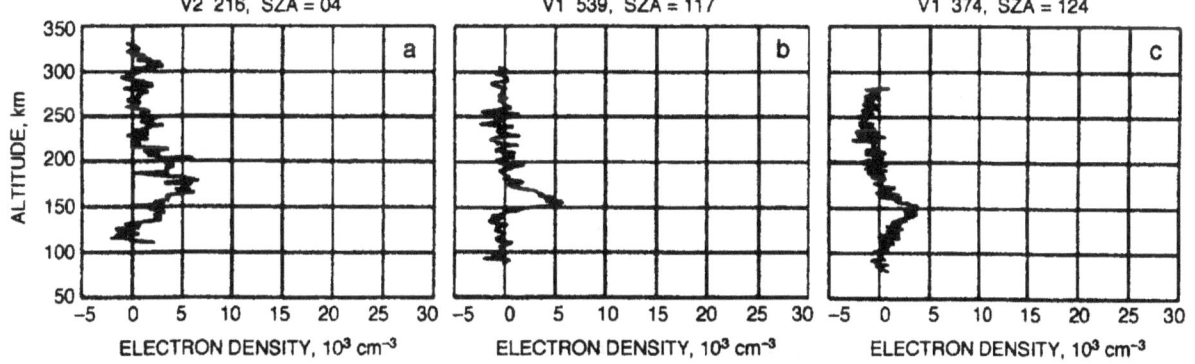

Figure 2-7. Three Examples of Mars Nightside Ionosphere Electron Density Profiles at Different Solar-Zenith Angles.

The nightside ionosphere is highly structured and dynamic, and low levels of plasma density may extend out a long distance in a comet-like structure [Vaisberg and Smirnov, 1986]. In a manner similar to the Venus's nightside ionosphere, we expect that the Martian nightside ionosphere also has attached and detached plasma clouds in the terminator regions and the ionosphere has some rays, density holes, and filament structures in the anti-sunward tail region [Gringauz, 1976a and b]. The nightward plasma flow from dayside across the terminators is the main source of the nightside ionosphere due to the large day-to-night pressure gradients. This cross-terminator flow strongly depends on solar wind pressure. During the solar minimum, the transport source is cut off. However, near solar maximum, high solar wind pressure can also sometimes suppress the altitude of dayside ionopause over the terminator, reducing the normal nightward transport. Horizontal transport from dayside is possible because Mars has almost no intrinsic magnetic field, which could inhibit plasma flow across the terminator. Local ion production by energetic electron impact is another important source. However, because of the lack of solar wind data during Mars nightside ionospheric observations, we do not know which mechanism is more important.

2.5   Ionospheric Effects on Radio Wave Propagation

We have plotted the dayside ionospheric density profiles for various values of SZA (χ) in Figure 2-8, using an average Martian dayside ionospheric model (Equation 2-1) and the following parameters: $N_0 = 2\times10^5$ cm^{-3}, $h_0 = 125$ km, $H = 11$ km, $k = 0.57$. The F$_2$ layer of Earth's ionosphere has a peak density of 2×10^6 cm^{-3} (2×10^{12} m^{-3}) at dayside and a peak altitude of ~300 km. The Martian ionospheric plasma density is one order of magnitude lower than Earth's, as shown in Table 2-1. The Martian dayside ionosphere at solar maximum has a peak density similar to that of the Earth nightside ionosphere at solar minimum. By integrating the dayside ionospheric profile, the total vertical electron content (TEC) of 4.0×10^{11}/cm^2 is obtained. This value is 50 times lower than Earth's ionospheric TEC. Even though the Martian ionospheric peak density and TEC are lower than in the Earth's ionosphere, we can still use them for ionospheric communication.

The Martian ionosphere can definitely be useful in future Mars ground-to-ground low-frequency communication. The Mars ionosphere can be used to perform trans-horizon (or beyond line of sight) communication for future Martian colonies, vehicles, and robots released from Mars landers. The Martian ionospheric critical frequency, $f_0 (MHz) = 9.0\times10^{-6}\sqrt{N_0(m^{-3})}$, is ~4.0 MHz for vertical incidence, which is a factor of 3 less than Earth's ionospheric critical frequency. For an oblique incident wave, the usable critical frequency $f_0 = 4.0$ (MHz) / cos θ_0, where θ_0 is the initial wave launch angle. Usable critical frequencies and single-hop distances, as a function of launch angle θ_0, are listed in Table 2-2. Hop distance (l) is a function of wave launch angle and ionospheric height (h) $l = 2\,h\,tan\,\theta_0$. When θ_0 increases, the maximum usable critical frequencies for oblique propagation increase significantly. The frequency is high enough to carry useful information. Figure 2-9 shows schematically how the Martian ionosphere can be used as a reflector for global communication. As noted earlier, the prevalent Martian sand storms have significant effects on the ionospheric peak height. For example, the 1971 global dust storm increased the peak height by 20–30 km. An increase in peak height (Δh) will cause an increase in hop distance. This is equivalent to an enhancement in hop distance by $\Delta l = 2\,\Delta h\,tan\,\theta_0$.

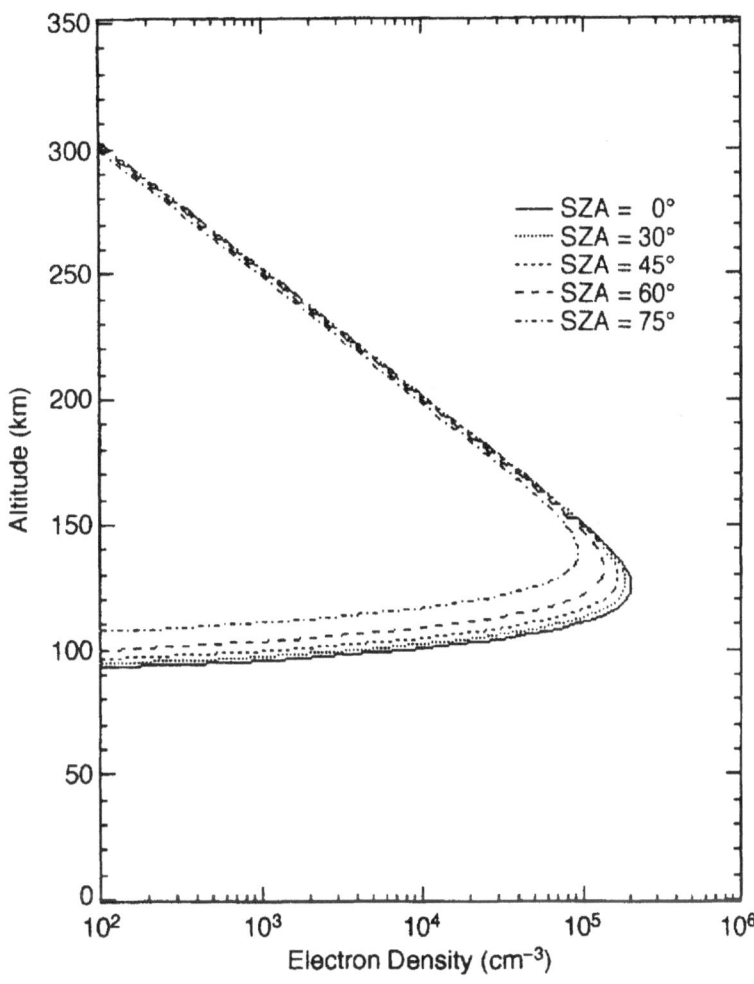

Figure 2-8. The Calculated Martian Dayside Ionospheric Altitude Profiles for Different Solar-Zenith Angles Using Equation 2-1 (Chapman layer model).

Table 2-1. Ionospheric Peak Density and Critical Frequency for Mars and Earth

Ionospheric Condition		Mars		Earth	
		n_0 (cm^{-3})	f_0 (MHz)	n_0 (cm^{-3})	f_0 (MHz)
Dayside	Solar Max.	2.5×10^5	4.5	2.0×10^6	12.7
	Solar Min.	1.0×10^5	2.9	5.0×10^5	6.3
Nightside*	Solar Min.	5.0×10^3	0.6	2.0×10^5	4.0
Dayside	TEC	4.0×10^{11} cm^{-2}		2.0×10^{13} cm^{-2}	

* There is no nightside ionospheric data available for Mars during solar maximum.

Table 2-2. Usable Critical Frequency and Hop Distance for Various Launch Angles

	Launch Angle θ_0					
	0°	**15°**	**30°**	**45°**	**60°**	**75°**
Maximum Usable Frequency (MHz)	4.0	4.14	4.62	5.66	8.0	15.5
One Hop Distance (km)	0	67.0	144.3	250.0	433.0	933.0

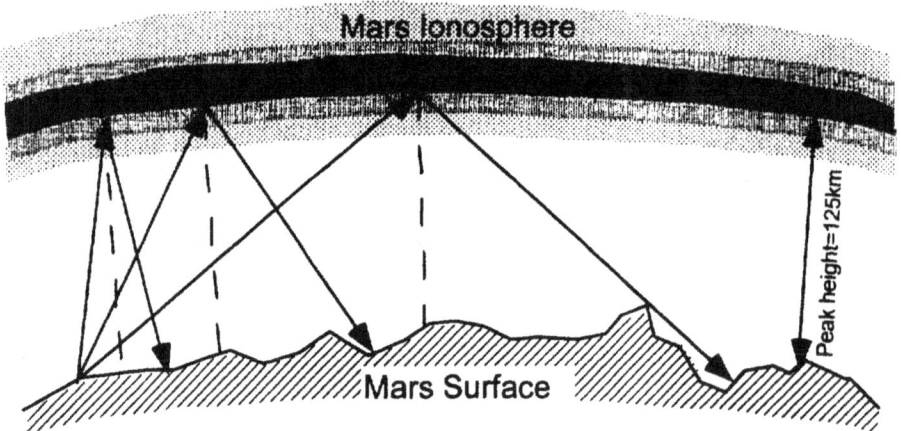

Figure 2-9. The Dayside Martian Ionosphere May Be Used as a Reflector for Trans-horizon Surface-to-Surface Communication. The maximum hop distance and usable frequency depend on the ray launch angles. Using this technique, one can keep in touch with areas beyond the line of sight.

The Martian dayside ionosphere has a stable peak density and peak height. During the solar maximum, N_0 can reach 2.5×10^5 cm^{-3} (4.5 MHz in f_0), while at solar minimum, there is a N_0 of $\sim 1.0 \times 10^5$ cm^{-3} (2.85 MHz in f_0). The ionospheric peak height is between 120 and 130 km. The stable condition is favorable to oblique incident communication using the ionosphere as a reflector from Martian surface to surface. The nightside ionosphere has very low plasma density ($\sim 5 \times 10^3$ cm^{-3}, corresponding to 0.63 MHz in f_0). A model for the nightside ionosphere is not available yet. More measurements and analyses need to be done for the nightside ionosphere under different solar conditions. The nightside ionospheric profile often shows no dominant density peak and has large variations. Because of the low usable frequency and very unstable condition, the nightside ionosphere is unlikely to be useful for global communication.

Because of the relatively low maximum density, based on the term $X = (f_0 / f)^2$ in Equation (1-1), the Martian ionospheric refractive index only significantly affects low frequency waves with frequencies below 450 MHz. The Martian ionosphere may slightly affect UHF links between a Mars lander (or rover) and an orbiter. Below 4.5 MHz, the cut-off frequency, the waves cannot pass through the ionosphere. The Martian ionosphere is almost transparent for radio waves with frequencies above 450 MHz. Because Mars has very little magnetic field ($\sim$50 nT), the term $Y =$

(f_B / f) also has little effect to the radio waves. The average gyrofrequency is about 1.5 kHz. The gyrofrequency has almost no effect on waves above 1 MHz.

The plasma collision frequency in the low ionosphere, ν, remains unknown, but it should depend on both plasma and neutral temperature and density. This frequency is estimated to be below 1.0 kHz. This term can cause low frequency wave attenuation within the ionosphere through the image part of $Z = (\nu/f)$. The scintillation effects of the Martian ionosphere on radio waves are not yet known. We expect that there are more plasma irregularities in the nightside ionosphere. Such irregularities can cause fluctuations in both amplitude and phase for low-frequency radio signals.

The effects of the Martian ionospheric total electron content (TEC = 4×10^{11}/cm^2) on radio wave propagation for vertical (elevation angles, $\psi = 90°$) one-way paths are summarized in Table 2-3. These effects include Faraday rotation, range delay, time delay, and phase advance. The Faraday rotation effect is near zero for a radial propagation wave because in most regions the draped magnetic field lies in the horizontal direction and B_L ($B_L = B_0 \cos \theta_{Bk}$) is near zero ($\theta_{Bk} = 90°$). Only in the nightside anti-solar region, where the magnetic fields are toward the ionotail, is there a larger rotation effect, as shown in the table. However, compared with Earth's ionosphere, all of these effects listed in the table are small because of the small TEC at Mars. We have also calculated signal Doppler frequency due to TEC changes and bandwidth coherence. Both of these effects are very small and can be neglected. For other elevation angles ($30°< \psi < 90°$), the TECs can be obtained by dividing the zenith TEC by $\sin \psi$. Thus, their ionospheric effects will also increase by $1/ \sin \psi$.

Table 2-3. Effects of the Total Vertical Electron Content of the Mars Ionosphere on Wave Propagation Characters (One-Way Path) for TEC=4×10^{11}/cm^2

	100 MHz	500 MHz	1 GHz	5 GHz	10 GHz
Faraday Rotation $\phi=(2.36\times10^4/f^2)B_L$ TEC	100"	4"	1"	0.04"	0.01"
Range Delay $\Delta R=(40.3/f^2)$TEC	16 m	0.66 m	0.16 m	0.007 m	0.002 m
Phase Advance $\Delta\phi=(8.44\times10^{-7}/f)$TEC	34 rad	7 rad	3.4 rad	0.7 rad	0.34 rad
Time Delay $\Delta t=(1.34\times10^{-7}/f^2)$TEC	54 ns	2.1 ns	0.54 ns	0.021 ns	0.005 ns
Doppler Frequency $f_D=1.34\times10^{-7}\Delta$TEC/$f\Delta t$*	3.7×10^{-3} Hz	1.9×10^{-3} Hz	3.7×10^{-4} Hz	1.9×10^{-4} Hz	3.7×10^{-5} Hz
Bandwidth Coherence $\Delta\phi/\Delta f=(-8.4\times10^{-7}/f^2)$TEC	-0.34×10^{-6} rad/Hz	-0.7×10^{-8} rad/Hz	-0.34×10^{-8} rad/Hz	-0.7×10^{-10} rad/Hz	-0.34×10^{-10} rad/Hz

*Assuming ΔTEC/$\Delta t = 1\times10^{16}$m^{-2}/hour.

2.6   Summary and Recommendations

The dayside ionosphere of Mars is generated through the photo-ionization of its upper atmosphere. The height of the ionosphere (ionopause) is dependent on solar wind pressure. A comet-like structure with low electron density extends several thousand kilometers at nightside.

The Martian dayside ionosphere may be described using a simple Chapman layer model. The ionosphere has a stable peak height and peak density. Its peak height is between 120 and 130 km. Martian dust storms significantly increase the ionospheric peak height. On average, the dayside Martian ionospheric plasma density is one order of magnitude lower than Earth's. Its TEC value is 50 times lower than Earth's ionospheric TEC.

The plasma density in the nightside ionosphere is very low (5×10^3 cm^3). The nightward plasma flow from dayside crossing the terminators is the main plasma source for the nightside ionosphere. The nightside ionospheric profile often shows no dominant density peak and has large variations.

Recommendation: The Martian ionosphere may play an important role in future Mars ground-to-ground global communications. The Martian dayside ionospheric critical frequency is ~ 4.0 MHz for vertical incidence. This frequency is high enough to carry useful amounts of data. The stable conditions in the dayside ionosphere are favorable for oblique-incidence communication using the ionosphere as a reflector for Martian surface-to-surface communication. Usable critical frequency and hop distance for an oblique-incidence wave significantly increase with the wave launch angle. Using the Martian ionosphere, we can also perform trans-horizon (or beyond line-of-sight) long-range communication for future Martian colonies, rover vehicles, and robots released from Mars landers. However, the nightside ionosphere has some limitations for global communication because of its low usable frequency and very unstable conditions.

The Martian ionosphere only effects the low frequency waves with frequencies below 450 MHz. Below 4.5 MHz, the cut-off frequency, the waves cannot pass through the ionosphere. It is almost transparent for radio waves with frequencies above 450 MHz. Because Mars has very little magnetic field (~ 50 nT), its gyrofrequency (f_B) has almost no effect on waves with frequencies above 1 MHz.

References

Acuna, M.H., et al., Magnetic field and plasma observations at Mars: Initial results of the Mars Global Surveyor Mission, *Science, 279*, 1676, 1998.

Bauer, S.J. and M.H. Hantsch, Solar cycle variation of the upper atmosphere temperature of Mars, *Geophys. Res. Lett., 16*, 373, 1989.

Bauer, S.J., and R.E. Hartle, On the extent of the Martian ionosphere, *J. Geophys. Res., 78*, 3169, 1973.

Chen, R.H., et al., The Martian ionosphere in light of the Viking observations, *J. Geophys. Res., 83*, 3871, 1978.

Cloutier, P.A., et al., Modification of the Martian ionosphere by the solar wind, *J. Geophys. Res.*, *74*, 6215, 1969.

Fox, J. L. and A. Dalgarno, Ionization, luminosity, and heating of the upper atmosphere of Mars, *J. Geophys. Res.*, *84*, 7315, 1979.

Gringauz, K.I., Interaction of solar wind with Mars as seen by charged particle traps on Mars 2, 3, and 5 satellite, *Rev. Geophys.*, *14*, 391, 1976a.

Gringauz, K.I., On the electron and ion components of plasma in the antisolar part of near-Martian space, *J. Geophys. Res.*, *81*, 3349, 1976b.

Hantsch, M.H. and S.J. Bauer, Solar control of the Mars Ionosphere, *Planet Space Sci.*, *38*, 539, 1990.

Hanson, W.B., and G.P. Mantas, Viking electron temperature measurements: Evidence for a magnetic field in the Martian ionosphere, *J. Geophys. Res.*, *93*, 7538, 1988.

Hanson, W.B., S. Sanatani, and D.R. Zuccaro, The Martian ionosphere as observed by the Viking retarding potentail analyzers, *J. Geophys. Res.*, *82*, 4351, 1977.

Kliore, A.J. et al, Radio occultation measurement of the Martian atmosphere over two regions by the Mariner IV space probe, *Moon and planets*, Edited by Dollfus, A., 226, North-Holand, Amsterdam, 1966.

Kliore, A.J., et al., Mariners 6 and 7: Radio occultation measurements of the atmosphere of Mars, *Science, 166*, 1393, 1969.

Kliore, A.J., D.L. Cain, G. Fjeldbo, B.L. Seidel, M.J. Sykes, and S.I. Rasool, The atmosphere of Mars from Mariner 9 radio occultation measurements, *Icarus, 17*, 484, 1972.

Kliore, A.J., et al., S Band radio occultation measurements of the atmosphere and topography of Mars with Mariner 9: Extended mission coverage of polar and intermediate latitude, *J. Geophys. Res.*, *78*, 4331, 1973.

Kolosov, M.A., Preliminary results of radio occultation studies of Mars by means of the Orbiter Mars 2, *Dokl. Akad. Nauk SSSR, 206*, 1071, 1972.

Luhmann, J.G., et al., Characteristics of the Mars-like limit of the Venus-solar wind interaction, *J. Geophys. Res .*, *92*, 8545, 1987.

Luhmann, J.G., et al., The intrinsic magnetic field and solar-wind interaction of Mars, in *Mars*, edited by H.H. Kieffer et al., The University of Arizona Press, Tucson & London, 1090, 1992.

Mantas, G.P., and W.B. Hanson, Analysis of Martian ionosphere and solar wind electron gas data from the planar retarding potential analyzer on the Viking spacecraft, *J. Geophys. Res.*, *92*, 8559, 1987.

McElroy, M.B., et al., Photochemistry and evolution of Mars' atmosphere: A Viking perspective, *J. Geophys. Res.*, *82*, 4379, 1977.

Nier, A.O., and M.B. McElroy, Composition and structure of Mars' upper atmosphere: Results from the neutral mass spectrometers on Viking 1 and 2, *J. Geophys. Res.*, *82*, 4341, 1977.

Ratcliffe, J.A., *An Introduction to the Ionosphere and Magnetosphere*, Cambridge University Press, New York, 1972.

Russell, C.T., et al., The magnetic field of Mars: Implications from gas dynamic modeling, *J. Geophys. Res., 98*, 2997, 1984.

Shinagawa, H. and T.E. Cravens, A one-dimensional multispecies magnetohydrodynamic model of the dayside ionosphere of Mars, *J. Geophys. Res., 94*, 6506, 1989.

Slavin, J.A., and R.E. Holzer, The solar wind interaction with Mars revisited, *J. Geophys. Res., 87*, 10285, 1982.

Slavin, J.A., et al., The Martian magnetosphere: Phobos 2 observations, *J. Geophys. Res., 96*, 11235, 1991.

Stewart, A.I., and W.B. Hanson, Mars upper atmosphere: Mean and variations, *the Mars reference atmosphere*, Ed. by Kliore, A.J., *Adv. Space Res., 2*, 2, 1982.

Vaisberg, O.L., Mars-plasma environments, in *Physics of Solar Planetary Environments*, Edited by D.J. Williams, 854, AGU, Washington, D.C., 1976.

Vaisberge, O., and V. Smirnov, The Martian magnetotail, *Adv. Space Res., 6*, 301, 1986.

Woo, R., and A. Kliore, Magnetization of the ionosphere of Venus and Mars, *J. Geophys. Res., 96*, 11073, 1991.

Zhang, M.H.G., et al., A post-Pioneer Venus reassessment of the Martian dayside ionosphere as observed by radio occultation methods, *J. Geophys. Res., 95*, 14829, 1990a.

Zhang, M.H.G., J.G. Luhmann, and A.J. Kliore, An observational study of the nightside ionospheres of Mars and Venus with radio occultation methods, *J. Geophys. Res., 95*, 17095, 1990b.

Zhang, T.L., et al., Magnetic barrier at Venus, *J. Geophys. Res., 96*, 11145, 1991.

3. Martian Atmosphere and Its Effects on Propagation

3.1 Introduction

When high-frequency radio waves pass through the Martian atmosphere, the signals also experience attenuation and impairment as they do in Earth's atmosphere. The signal degradation mainly takes place in the lower part of the atmosphere: the troposphere. The propagation medium in the Martian troposphere includes gases, water vapor, cloud, fog, ice, dust and aerosols (haze), etc. The impairment mechanisms include absorption, scattering, refraction, diffraction, multipath, scintillation, Doppler shift, etc. Impairment phenomena include fading, attenuation, depolarization, frequency broadening, ray bending, etc. To measure the degradation of the signals, the following observable parameters are usually used: amplitude, phase, polarization, frequency, bandwidth, and angle of arrival. However, compared with Earth's atmosphere, the Martian atmosphere is thin. Thus, we expect a much smaller tropospheric effect on radio wave propagation.

Before we study radio wave propagation in the Martian atmosphere, it is interesting to calculate the speed of sound v_s at the Martian surface

$$v_s = \sqrt{\frac{\gamma P_0}{\rho_0}}$$

(3-1)

where γ is adiabatic index of the gases (~1.35), P_0 is surface atmospheric pressure (6.1 mb), and ρ_0 is surface atmospheric mass density (0.02 kg/m^3). Thus, the speed of sound at Mars is 206 m/s, while the speed at Earth is 331 m/s. Actually, the Martian atmosphere is so much thinner that it is hard for a sound wave to propagate. Consequently, it will be necessary for future Martian colonists to use radio wave communications.

Most Martian tropospheric parameters are those studied in Martian meteorology, such as its pressure, temperature, and wind. Based on measurements from the Vikings and from Mars 6, a nominal mean model for pressure and temperature was obtained [Seiff, 1982]. Table 3-1 lists temperature, pressure, and atmospheric mass density as a function of altitude with 2-km increments for a daily summer-seasonal mean mid-latitude atmosphere. Table 3-1 also gives the ratio p/p_0, as well as pressure values, so that the model may be applied to other values of p_0 than those selected for other times in the summer seasons. The model has the following equation form:

$$\frac{p}{p_0} = \exp\left[-\frac{\mu}{R}\int_{z_0}^{z}\frac{g(z)}{T(z)}\,dz\right]$$

(3-2)

Table 3-1. Nominal Summer-Seasonal Mid-Latitude Martian Atmospheric Model

z, km	T, K	p/p_0	Northern Summer			Southern Summer	
			p, mb p_0 = 6.36	ρ, kg/m³	g, m/s²	p, mb p_0 = 7.3	ρ, kg/m³
0	214	1.000	6.36	1.56×10^{-2}	3.730	7.30	1.78×10^{-2}
2	213.8	0.833	5.30	1.30	3.725	6.08	1.49
4	213.4	0.694	4.41	1.08	3.720	5.07	1.24
6	212.4	0.579	3.68	9.07×10^{-3}	3.716	4.23	1.04
8	209.2	0.481	3.06	7.65	3.712	3.51	8.78×10^{-3}
10	205	0.399	2.54	6.47	3.708	2.91	7.42
12	201.4	0.330	2.10	5.45	3.703	2.41	6.25
14	197.8	0.2715	1.73	4.57	3.699	1.98	5.24
16	194.6	0.2229	1.42	3.81	3.695	1.63	4.37
18	191.4	0.1825	1.16	3.17	3.690	1.33	3.64
20	188.2	0.1489	9.47×10^{-1}	2.63	3.686	1.09	3.02
22	185.2	0.1211	7.70	2.18	3.686	8.84×10^{-1}	2.50
24	182.5	9.82×10^{-2}	6.25	1.79	3.678	7.17	2.05
26	180	7.95	5.06	1.47	3.673	5.81	1.69
28	177.5	6.41	4.08	1.20	3.669	4.68	1.38
30	175	5.16	3.28	9.81×10^{-4}	3.664	3.76	1.13
32	172.5	4.14	2.63	7.98	3.660	3.02	9.16×10^{-4}
34	170	3.31	2.11	6.48	3.656	2.42	7.44
36	167.5	2.637	1.68	5.24	3.652	1.93	6.01
38	164.8	2.095	1.33	4.23	3.648	1.53	4.86
40	162.4	1.660	1.06	3.40	3.643	1.22	3.90
42	160	1.310	8.33×10^{-2}	2.72	3.638	9.56×10^{-2}	3.12
44	158	1.032	6.56	2.17	3.634	7.53	2.49
46	156	8.10×10^{-3}	5.15	1.73	3.630	5.91	1.99
48	154.1	6.34	4.03	1.37	3.626	4.63	1.57
50	152.2	4.95	3.15	1.08	3.622	3.62	1.24
52	150.3	3.86	2.45	8.54×10^{-5}	3.618	2.81	9.80×10^{-5}
54	148.7	2.99	1.90	6.69	3.614	2.18	7.68
56	147.2	2.319	1.47	5.24	3.609	1.69	6.01
58	145.7	1.792	1.14	4.09	3.605	1.31	4.69
60	144.2	1.382	8.79×10^{-3}	3.19	3.601	1.01	3.66
62	143	1.063	6.76	2.47	3.597	7.76×10^{-3}	2.84
64	142	8.17×10^{-4}	5.20	1.91	3.592	5.97	2.19
66	141	6.26	3.98	1.48	3.588	4.57	1.70
68	140	4.79	3.04	1.14	3.584	3.49	1.31
70	139.5	3.67	2.33	8.75×10^{-6}	3.580	2.67	1.00
72	139	2.80	1.78	6.70	3.576	2.04	7.69×10^{-6}
74	139	2.141	1.36	5.12	3.572	1.56	5.88
76	139	1.637	1.04	3.92	3.568	1.19	4.50
78	139	1.252	7.96×10^{-4}	3.00	3.564	9.14×10^{-4}	3.44
80	139	9.57×10^{-5}	6.09	2.29	3.559	6.99	2.63
82	139	7.32	4.66	1.75	3.555	5.35	2.01
84	139	5.61	3.57	1.34	3.551	4.10	1.54
86	139	4.29	2.73	1.03	3.547	3.13	1.18
88	139	3.29	2.09	7.87×10^{-7}	3.543	2.40	9.03×10^{-7}
90	139	2.518	1.60	6.03	3.539	1.84	6.92
92	139	1.929	1.23	4.62	3.535	1.41	5.30
94	139	1.479	9.41×10^{-5}	3.54	3.531	1.08	4.06
96	139	1.134	7.21	2.71	3.527	8.28×10^{-5}	3.11
98	139	8.70×10^{-6}	5.53	2.08	3.523	6.35	2.39
100	139	6.67	4.24	1.60	3.519	4.87	1.84

where $\mu = 43.49$, $R = 191.18$ joul/kg K, gravity $g(z)$, and temperature $T(z)$ values at altitude z can also be found in Table 3-1. A plot of the modeled pressure profile is given in Figure 3-1. The pressure decreases in the model by five orders of magnitude from the surface up to 100 km.

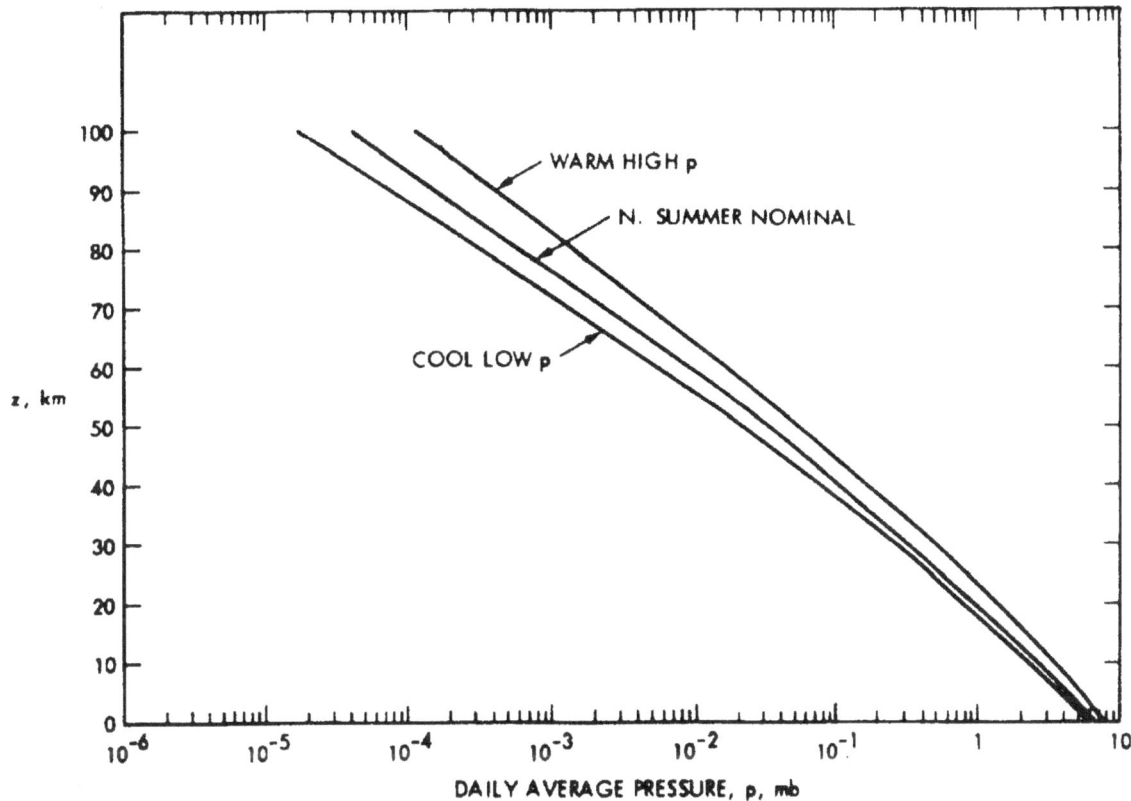

Figure 3-1. Nominal Northern Summer Mid-Latitude Model of the Atmosphere of Mars and Variation of Warm and Cool Summer (from Seiff, 1982).

3.2 Martian Tropospheric Effects

The Martian atmospheric refractive index governs the propagation of radio waves. The index is a function of the atmospheric pressure and temperature, as shown in Equation (1-3). To understand variations of the index, we need to study both Martian atmospheric pressure and temperature. The atmosphere of Mars is much thinner than that of Earth, with a surface pressure averaging 1/100th that at the surface of the Earth. Barometric pressure varies at each landing site on a semi-annual basis. When the southern cap is largest, the mean daily pressure observed by Viking Lander 1 was as low as 6.8 millibars (mb); at other times of the year, the observed pressure was as high as 9.0 mb. The pressures at the Viking Lander 2 site were 7.3 and 10.8 mb (730 and 1080 Pa). In comparison, the average pressure of the Earth is 1013 mb (1.013×10^5 Pa).

Figure 3-2 shows one of the latest atmospheric pressure profiles measured by MGS occultation from the Martian surface to 45-km altitude [Hinson et al., 1999]. The pressure was measured at the local time of 0535 a.m. and in late fall. In this particular instance the pressure was 6.3 mb at the 0-km altitude, even though the 0-km altitude is defined as the reference surface where the atmospheric pressure averages 6.1 mb (610 Pa); dust storms often increase the atmospheric pressure. In November 1997, MGS observed the thermospheric response to the dust storm, increasing the altitude of the thermospheric pressure surface by 8 km at middle north latitude [Keating et al., 1998].

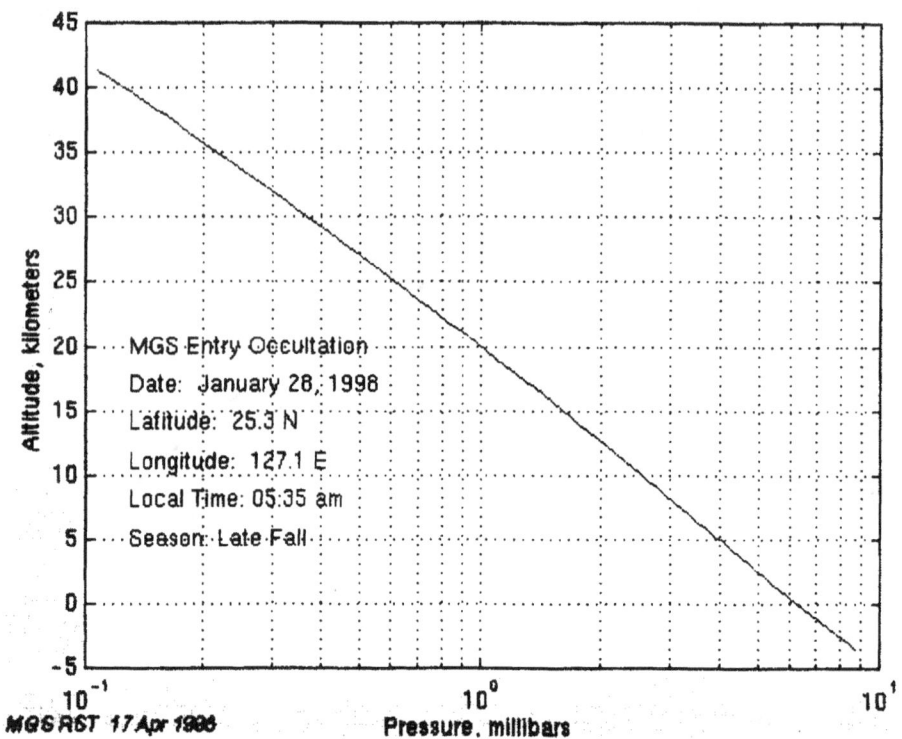

Figure 3-2. Atmospheric Pressure Profile Measured by MGS Radio Occultation on January 28, 1998.

Figure 3-3 shows, respectively, altitude profiles of the atmospheric density (left) and temperatures (right) measured by Mars Pathfinder as it descended through the nighttime atmosphere and landed on the surface [Schofield et al., 1997]. Figure 3-3 shows that near the surface where Pathfinder landed (at night) the temperature was 200 K, in the middle of the atmosphere the temperature was 100 K, and at the uppermost reaches of the atmosphere the temperatures ranged from 150 to 300 K. This uppermost region of the atmosphere is called the exosphere.

Surface temperatures range from 140 K (–133°C) at the winter pole to 296 K (23°C) on the dayside equator during summer. Although the length of the Martian day (24 hours and 37 minutes) and the tilt of its axis (25 degrees) are similar to those on Earth (24 hours and 23.5 degrees), the orbit shape of Mars around the Sun affects the lengths of the seasons the most. Figure 3-4 shows one of atmospheric temperature profiles from MGS occultation measurements [Hinson et al., 1999]. The temperature has a peak value of 218 K (–55°C) at the 10-km altitude.

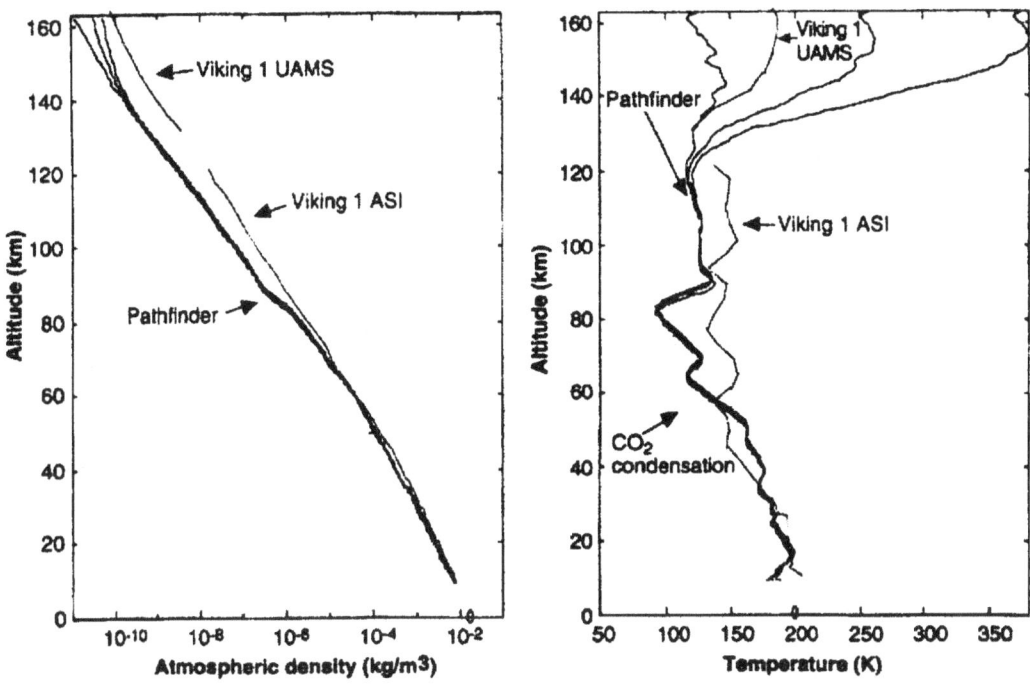

Figure 3-3. (Left) The Atmospheric Density Profiles Derived from the Mars Pathfinder
Accelerometer Data. Results from the VL-1 atmospheric structure instrument (ASI)
and the Viking 1 upper atmosphere mass spectrometer (UAMS) are also plotted for
comparison. (Right) The atmospheric temperature profiles derived from the Pathfinder
measurements and from the VL-1 ASI, UAMS experiments, and the CO_2 condensation.
The surface density and temperature measured by the Pathfinder MET instrument
(oval) are also shown.

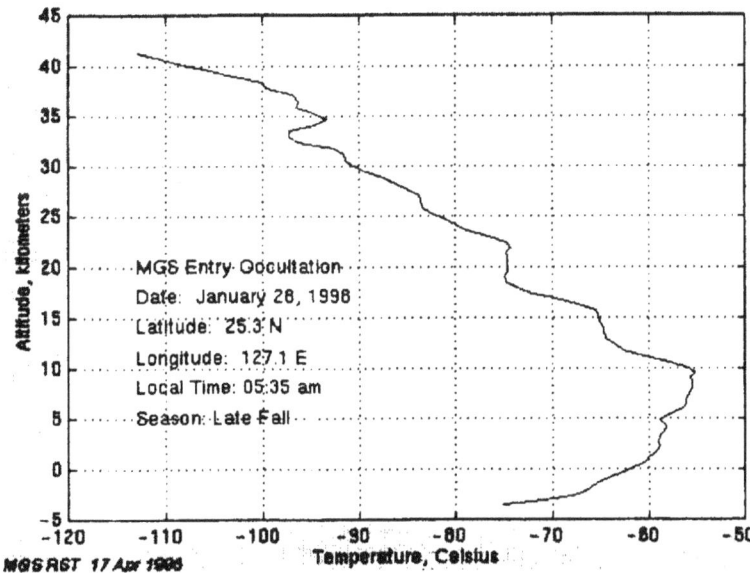

Figure 3-4. Martian Atmospheric Temperature Profile Measured by MGS Radio
Occultation on January 28, 1998.

25

Above an altitude of 10 km, the temperature decreases with height. This is the usual behavior in atmospheres or atmospheric layers. Below 10 km, however, the temperature increases with height. This is called a temperature nighttime inversion. In this case, the inversion results from the radiation of infrared energy from the surface of Mars and the atmosphere in close contact with it, which occurs through the night hours [Hinson et al., 1999]. The loss of energy leads to cooling. The same phenomenon takes place here on Earth on clear nights with little or no wind. Radiation inversions generally dissipate in the hours after the Sun rises and the surface is warmed once again.

The Viking Lander meteorological sensors gave detailed information about the atmosphere. They found patterns of diurnal and longer-term pressure and temperature fluctuations. The temperature reached its maximum of 238 K every day at 2 p.m. local solar time and its minimum of 190 K just before sunrise. Local time changes in the temperature profiles in the lowest 8 km are modeled in Figure 3-5. In the lowest 4 km, there is a boundary layer that is strongly influenced by radiative exchange with the ground and in which an intense nighttime inversion forms. On Mars, the atmospheric pressure varies with the seasons. During winter, it is so cold that 20 to 30 percent of the entire atmosphere freezes out at the pole, forming a huge pile of solid carbon dioxide. The pressure minimum of just under 6.7 mb (roughly 0.67 percent of pressure at Earth sea level) was reached on sol 100, the 100th Martian day after the Viking 2 landing, as shown in Figure 3-6. The pressure minimum seen by Pathfinder indicates that the atmosphere was at its thinnest (and the south polar cap probably its largest) on sol 20.

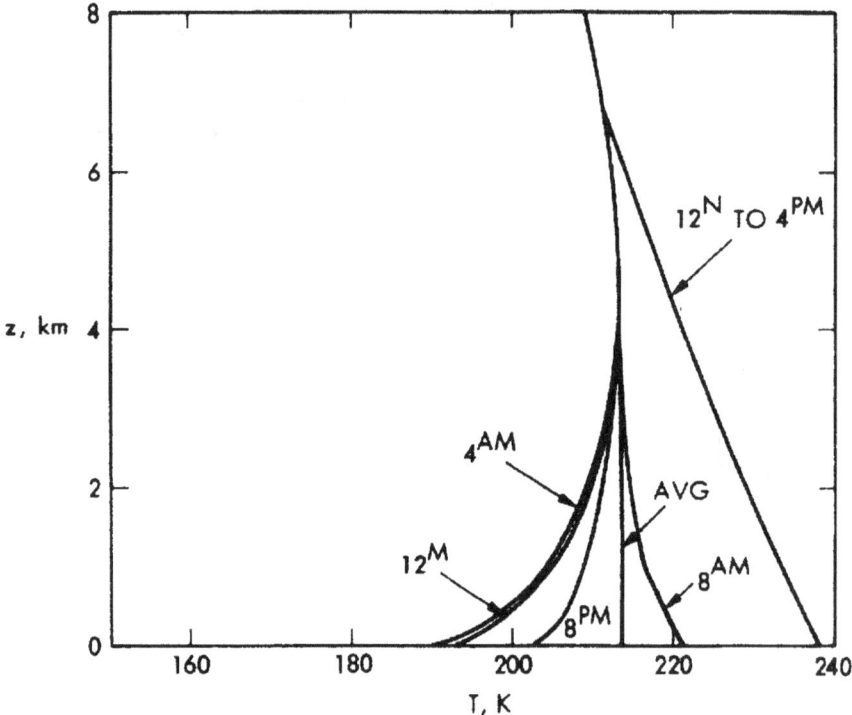

Figure 3-5. Models of Martian Atmospheric Surface Temperature Variation and Temperature Profiles in the Lowest 8 km (Seiff, 1982).

Morning temperatures fluctuated abruptly with time and height. The Viking sensors positioned at 0.25, 0.5, and 1 m above the surface obtained different readings. This suggests that cold morning air is warmed by the surface and rises in small eddies, or whirlpools; this is very different from what happens on Earth. Afternoon temperatures, after the atmosphere had warmed, did not show the same variations. In the early afternoon, dust devils repeatedly swept across the lander. They showed up as sharp, short-lived pressure changes, and they were probably similar to events detected by the Viking landers and orbiters. Such dust devils may be an important mechanism for raising dust into the Martian atmosphere. The prevailing winds were light (less than 10 m/s, or 36 km/hr) and variable.

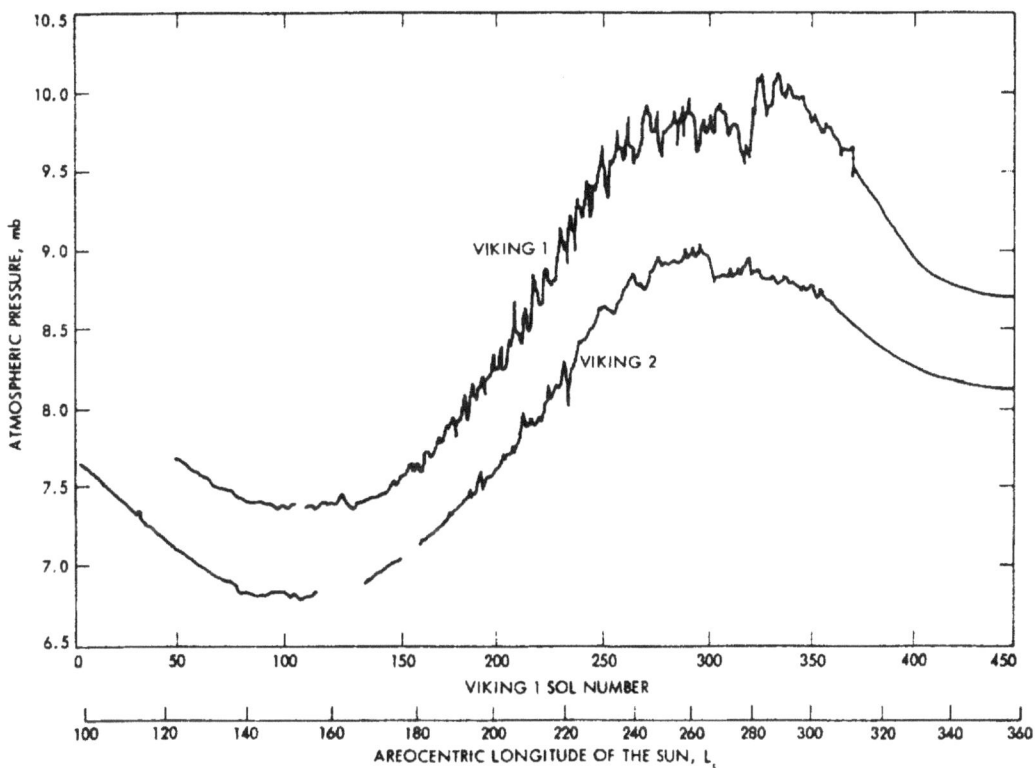

Figure 3-6. Seasonal Variation of Surface Pressure at the Two Viking Sites (Leovy, 1982).

Mars Pathfinder measured atmospheric conditions at higher altitudes during its descent. The upper atmosphere (above 60-km altitude) was colder than Viking had measured. This may simply reflect seasonal variations and the time of entry: Pathfinder came in at 3 a.m. local solar time, whereas Viking arrived at 4 p.m., when the atmosphere is naturally warmer. The lower atmosphere was similar to that measured by Viking, and its conditions can be attributed to dust mixed uniformly in comparatively warm air.

Two major effects of the Martian troposphere on radio wave propagation are probably multipath (due to the refraction) and scintillation (due to irregularities). Other effects in a clear Martian atmosphere (such as fading, dispersion, etc.) are neglected. Because Martian atmospheric pressure and temperature vary with altitude, vertical gradients exist in the refractive index in the troposphere (lower atmosphere). This causes multipath effects for radio waves with different launch angles due to ray bending. When a radio ray is not launched exactly vertically (perpendicular to the gradient), it gradually changes its direction, sometimes even going back to the Martian surface. Since the refractive index varies mainly with altitude, only the vertical gradient of the refractive index n [Bean and Dutton, 1966] is considered in most cases. The bending of a ray at a point is expressed by

$$\frac{1}{\rho} = \frac{\cos\varphi}{n}\frac{dn}{dh}$$

(3-3)

where ρ is radius of curvature, dn/dh is vertical gradient of the refractive index, h is height of the point above the Mars surface, and φ is the angle of the path relative to the horizontal of the point considered.

Only when a wave's launch angle is near $0°$ (close to horizontal) can the ray be trapped in a horizontal ducting layer. Because $n \approx 1$, from equation (3-3), we have an approximation

$$\frac{1}{\rho} = \frac{dn}{dh} = 10^{-6}\frac{dN}{dh}$$

(3-4)

where radio refractivity $N = (n-1) \times 10^6 = 130.6 P / T$ (N unit) from Equation (1-4), P is atmospheric pressure (in mb), and T is temperature (in Kelvins) for a dry atmosphere (Martian atmospheric water vapor pressure is neglected here). Based on the altitude profiles of Martian atmospheric P and T shown in Figures 3-4 and 3-5, we have calculated and plotted the altitude profile of the radio refractivity N in Figure 3-7. This profile can be fitted using a function

$$N(h) = N_0 \exp(-h / H_N)$$

(3-5)

where N_0 is the surface value of N when altitude $h = 0$, and H_N is the refractivity scale height. From the best fitting, we can find that $N_0 = 3.9$ (N unit) at Martian surface and $H_N = 11.0$ km. We also have

$$dN / dh = -N(h) / H_N$$

(3-6)

Because the Martian atmosphere has a very small N (only 1.2% of the Earth's, 315 N unit) and also a very small gradient dN/dh, there will be very small ray-bending effects. Only when the wave angle is very close ($\varphi < 0.3°$) to the horizon, can the wave ray be trapped by a horizontal duct (surface or elevated). Temperature changes near the surface have little effect on the N. For example, a 20°C change in temperature detected by Viking 1 and 2, as shown in Figure 3-8, only makes a contribution of 0.22 N unit (about 10%) to N. Even through the ray-bending effect is small, exact phase delays, range delays, and appearing angle deviations need to be calculated. When the wave propagates nearly horizontally and is trapped in a surface duct, it bounces back inside the waveguide between the Martian surface and the top reflection layer. Because Mars is only about half the size of Earth and because Mars has a larger surface curvature than Earth, it is expected that the signals will have a greater defocusing loss.

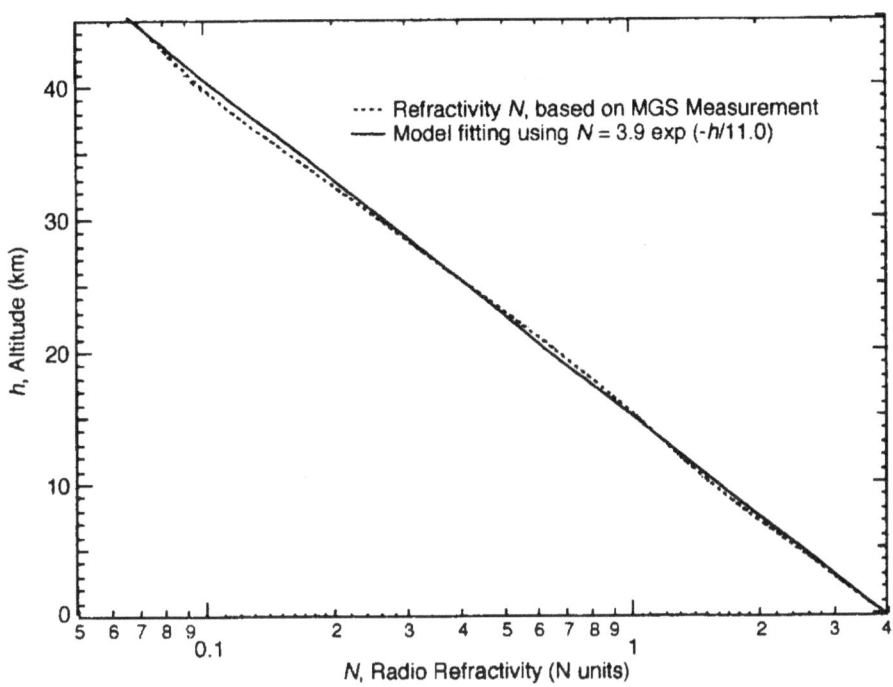

Figure 3-7. Radio Refractivity for Martian Atmosphere. Dry air pressure and temperature profiles are used for the refractivity calculation.

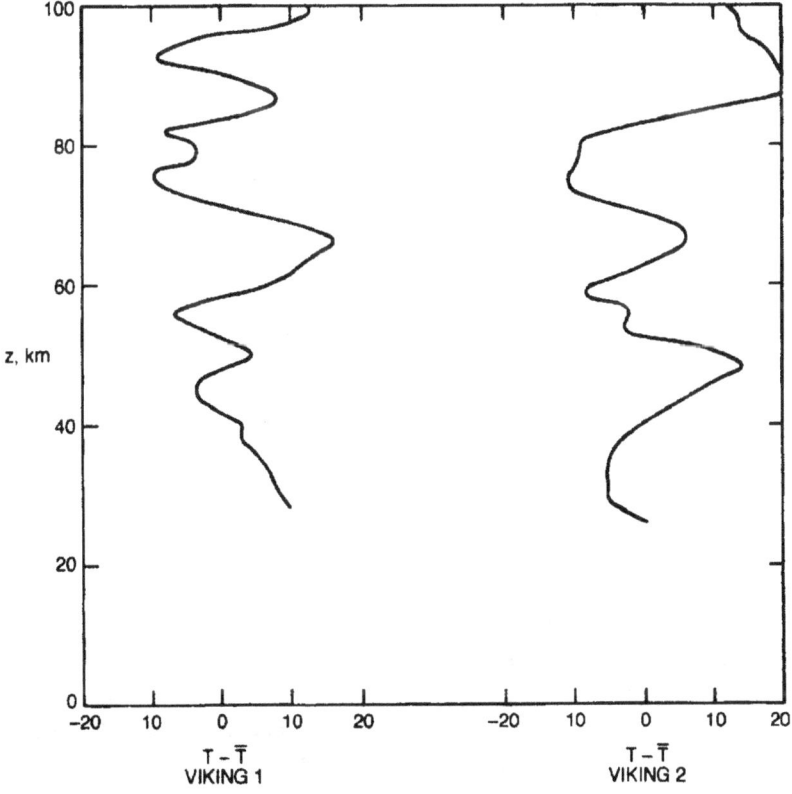

Figure 3-8. Temperature Oscillations Found in the Viking Temperature Soundings (from Seiff, 1982).

Tropospheric scintillation is caused by turbulence-induced spatial and temporal refractive index variation. Refractive index variations also cause wavefront distortions and increase the bit- error rate. Tropospheric scintillation caused by refractive index irregularities has been observed on Earth space paths at frequencies to above 30 GHz.

It is hard to determine from available Mars data the extent of Martian tropospheric turbulence and irregularities. However, refraction index theories developed for Earth's atmosphere can be applied to the Martian atmosphere. There is a relation between the size of the mean square fluctuations of the refractive index C_N and the size of the mean square fluctuations of temperature C_T [Annis, 1987]:

$$C_N \propto \frac{P}{T^2} C_T \qquad (3\text{-}7)$$

where T and P are temperature and pressure in the Martian troposphere. The ratio of Mars to Earth in fluctuations is

$$\frac{C_{N,Mars}}{C_{N,earth}} = \frac{P_{Mars}}{P_{Earth}} \left(\frac{T_{Earth}}{T_{Mars}} \right)^2 \frac{C_{T,Mars}}{C_{T,Earth}} \approx 0.5\% \frac{C_{T,Mars}}{C_{T,Earth}} \qquad (3\text{-}8)$$

Thus, refraction index variation fluctuations in Martian troposphere should be only about 0.5% of that in the Earth atmosphere, if C_T is the same for Mars and Earth.

3.3   Martian Clouds and Fogs

The Martian atmosphere contains only a very small amount of water (0.03% by volume, which is 1/300 to 1/1000 as much water as Earth). However, because of low pressure and temperature, the water can still condense out to form clouds in the atmosphere. Unlike the Earth, where clouds are found around the entire globe, on Mars, clouds seem to be plentiful mainly below the middle latitude region, as shown in the Hubble telescope image in Figure 3-9. Many of the cloud formations seem to be due to topographic forcing by Olympus Mons. This may be because water on Mars is mainly found around the equator and low latitudes. Recent detailed pictures from MGS have revealed many young gullies, possibly formed by flowing water, at the Martian surface between latitudes 30° and 70°. Because the Martian surface atmospheric pressure is so low, the water must be quickly evaporated or go underground [Malin and Edgett, 2000].

As early as 1796 scientists were reporting "yellow" clouds and "white" or "bluish" clouds in the Martian atmosphere. However, it wasn't until the Mariner 9 mission that clouds of water were positively identified. Mars Global Surveyor is providing more proof of the existence of water clouds. Using its thermal emission spectrometer, MGS detected water in some clouds. Mars Pathfinder took images of Martian clouds from the ground level. A few clouds have been seen at the north pole [Briggs et al., 1977]. This may have been because the north polar ice cap was evaporating with the coming of the northern spring season.

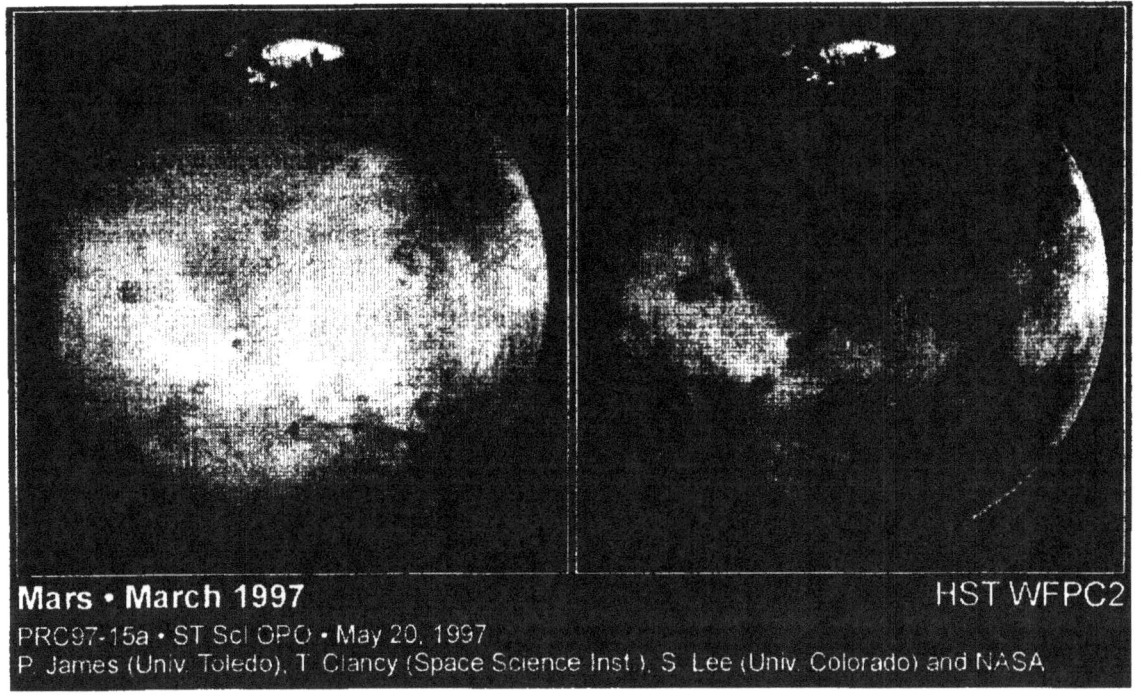

Figure 3-9. A Hubble Telescope Image of Martian Clouds. The clouds are found mostly in the equatorial and low latitude regions.

In the Earth atmosphere, rain droplets have larger particle sizes than water particles in clouds and fogs (< 0.01 cm). Thus, Rayleigh scattering theory applies. There is a general relation between the wave attenuation and total water content per unit volume for Rayleigh scattering.

$$\alpha_c = k_i \rho_l \tag{3-9}$$

where α_c is the specific attenuation (dB/km) within clouds, k_i is the attenuation coefficient (dB/km/gm/m^3) determined from Rayleigh scattering theory, and ρ_l is water content (g/m^3). The total cloud attenuation can be obtained by computing the total content of water along the path. Based on Rayleigh scattering, the coefficient k_i is a function of dielectric permittivity ε and relative dielectric permittivity K_c for frequencies up to 100 GHz, as shown below.

$$k_i = 0.4343 \frac{6\pi}{\lambda} Im\left[-\frac{K_c - 1}{K_c + 2}\right] \tag{3-10}$$

where λ is the wavelength, Im indicates the imaginary part of K_c, and K_c is the complex relative dielectric permittivity of water or ice (that is dielectric permittivity $\varepsilon = \varepsilon_0 K_c$, where ε_0 dielectric permittivity at free space). The quantity K_c is a function of temperature and frequency.

Cloud drop sizes, liquid water content, relative dielectric permittivity K_c, and the coefficient k_i for the Earth atmosphere are well documented in the studies of Gunn and East [1954], Battan [1973], Ludlam [1980], and Slobin [1982]. Table 3-2 shows values of the imaginary part of $-(K_c - 1)/(K_c + 2)$, adapted from Battan [1973]. Values of attenuation coefficient k_i by water and ice clouds were calculated by Gunn and East [1954] for various wavelengths at various temperatures and are given in Table 3-3. These values can provide an upper limit for Martian cloud attenuation.

Table 3-2. Values for Complex Relative Dielectric Permittivity of Water Ice
$Im\ [-(K_c - 1)/(K_c + 2)]$ **(adapted from Battan, 1973)**

Substance	T (°C)	$\lambda = 10$ cm	$\lambda = 3.21$ cm
Water Ice	0°	9.6×10^{-4}	9.6×10^{-4}
Water Ice	−10°	3.2×10^{-4}	3.2×10^{-4}
Water Ice	−20°	2.2×10^{-4}	2.2×10^{-4}

Table 3-3. One-way Attenuation Coefficient, k_i in Clouds (dB/km/g/m^3)

Temperature K (°C)	Wavelength (cm)			
	0.9	1.24	1.8	3.2
Water Ice 273 (0)	8.74×10^{-3}	6.35×10^{-3}	4.36×10^{-3}	2.46×10^{-3}
Cloud 263 (−10)	2.91×10^{-3}	2.11×10^{-3}	1.46×10^{-3}	8.19×10^{-4}
253 (−20)	2.0×10^{-3}	1.45×10^{-3}	1.0×10^{-3}	5.63×10^{-4}

From the tables, we can see that attenuation decreases with increasing wavelength. For signals with frequencies of 10 GHz and lower, attenuation due to clouds is small. Values of $Im\ [-(K_c - 1)/(K_c + 2)]$ for ice clouds are independent of wavelengths.

At Earth, for signals with frequencies below 100 GHz, fog attenuation is not significant. Medium liquid fog typically has a water content of about 0.05 gm/m^3 and a visibility (V) of 300 m. This causes a 0.4-dB/km attenuation for a radio wave with a frequency of 140 GHz. The attenuation for a thick fog ($\rho_l = 0.5$ gm/m^3 and $V = 50$ m) is 4 dB/km. In additional to the cloud attenuation, clouds also can increase the system noise temperature because clouds are a source of emission as well as absorption [Slobin, 1982].

Because of the very low temperatures, Martian clouds probably consist of ice crystals. Some clouds may consist of CO_2 droplets. Radio wave attenuation due to ice clouds is two orders of magnitude smaller than that of water clouds, but water clouds have a strong depolarization effect on radio waves. For the Martian atmospheric cloud attenuation study, the problem is that at this stage we do not know what percentage of Martian clouds consist of water liquid or ice and what percentage are CO_2 clouds. We also do not have any direct measurement of water content within the clouds. An alternative way to calculate the scattering attenuation due to cloud and fog on radio waves is to use the observed optical depth through the following relation:

$$A(\lambda) = 54.62 \frac{r\tau}{\lambda} \left[\frac{3\varepsilon''}{(\varepsilon'+2)^2 + \varepsilon''^2} \right] \tag{3-11}$$

where $A(\lambda)$ is attenuation in dB/km, τ is optical depth, r is the particle radius in meters, λ is the wavelength in meters, and ε' and ε'' are the real and imaginary parts of the mean permittivity of the cloud droplets. The optical depth, τ, is a measure of attenuation over the entire path taken from the ground to space. Optical depth increases as the line of sight moves down toward the horizon because of increasing path length. The optical depth may be obtained through the following measurements. The power received, P_r, is the power transmitted, P_t, multiplied by the attenuation: $P_r = P_t e^{-\tau}$ (i.e., $\tau = ln(P_t / P_r)$).

There is a distinct seasonal dependence as well as a latitude dependence for Martian cloud distributions. A longitudinal dependence and a time-of-day dependence are not obvious. Some Martian clouds form at dawn and burn off rapidly, and others form only in midday. In general, the northern hemisphere is covered with more clouds than the southern hemisphere [Kahn, 1984]. Clouds are relatively abundant during northern spring and summer at mid-latitudes. In the southern hemisphere the situation is complicated by atmospheric dust. Overall, the Martian clouds have an optical depths of 0.05–3.0, a figure closer to terrestrial thin and high-level cirrus clouds ($\tau = 0.5$–3.5). For comparison, a terrestrial stratus cloud has an optical depth of ~ 6–80, while a cumulus cloud has an optical depth of ~ 5–200. It should be mentioned that these optical depths are at visual wavelengths (0.67 microns).

More study is needed to understand just how clouds are formed in the Martian atmosphere. For example, even though clouds have been found, we still do not know whether there is rain on Mars. Atmospheric temperatures reported by Mars Pathfinder during its descent indicate that it may be too cold in the cloud-forming region of the Martian atmosphere for droplets to fall to the ground as liquid, but it may be cold enough for the condensation of CO_2 droplets.

To answer these questions, 58,000 images of Mars provided from Viking Orbiters and Mariner 9 have been analyzed. Cloud distributions with seasons and latitudes have been obtained [Kahn, 1984; Thorpe, 1977]. Mars Pathfinder took measurements of many clouds in the Martian sky from the surface of Mars itself. The images of the Martian sky from the 80-day mission provided further assessment of Martian weather patterns.

Widespread thick clouds mainly occur in three regions: the polar hoods, the Tharsis bulge, and the plateau region in the southwestern end of the Marineris valley. These clouds usually have thinner optical depths ~0.05. Isolated clouds include the following:

1. Lee waves. These clouds form in the lee of large obstacles such as mountains, ridges, craters, and volcanoes.

2. Wave clouds. These clouds appear as rows of linear clouds. They are common at the edge of the polar caps.

3. Cloud streets. These clouds exhibit a double periodicity. They appear as linear rows of cumulus-like, bubble-shaped clouds.

4. Streaky clouds. These clouds have a preferred direction without periodicity.

5. Plumes. These are elongated clouds. They appear to have a source of rising material and in many cases are composed of dust particles.

As a summary, we list optical depths for both Earth and Mars cloud patterns in Table 3-4.

Table 3-4. Visual Optical Depths of Clouds and Fogs on Earth and Mars*

Atmospheric Condition	Earth		Mars	
	Optical Depth	Distribution	Optical Depth	Distribution
Clouds H_2O	~5	50% coverage	~1.0	Winter polar; behind high places
Clouds CO_2	None	None	~0.001 ~1.0	Many places Winter polar
Fog	~3	Many places	~0.2 ~1.0	Morning Valleys & crater bottoms
Aerosol Dust	Variable	Variable	0.5	Everywhere
Dust Storms	Variable	Variable	10.0	Southern Hemisphere or global

*Adapted from Annis [1987].

Martian fog usually occurs in low areas such as valleys, canyons, and craters. It forms during the coolest times of the day, such as dawn and dusk [Annis, 1987]. The fog seems to burn off in the afternoon. Fog seen by the Viking Landers was thin, about $\tau = 0.2$. Sometimes Martian ground haze is caused by dust in the atmosphere; however, if the atmosphere is clear, ground fog can be easily identified.

Using the optical depths listed in Table 3-4 and Equation (3-11) for a given droplet radius and permittivity, we can calculate the wave attenuation for various wavelengths. It is believed that the total attenuation for Ka-band radio wave signals is about 0.1 dB.

3.4 Martian Aerosols

It is believed that Martian dust particles are the main contribution to the Martian aerosols. The Martian sky appears pink and somewhat dark at sunset. This is because there are not enough molecules in the atmosphere to scatter the amount of light we are used to seeing on Earth. Also, the many rust-colored dust particles in the atmosphere contribute to the pink color.

Mars Pathfinder found that the sky on Mars had the same pale pink color as it did when imaged by the Viking landers [Golombek et al., 1997; Schofield et al., 1997]. Fine-grained, bright-red dust in the atmosphere would explain this color. This suggests that the Martian atmosphere always has some dust in it from local dust storms. The inferred dust-particle size (roughly a micron), shape, and amount of water vapor (equivalent to a meager one-hundredth of a millimeter of rainfall) in the atmosphere are also consistent with measurements made by Viking.

On Mars, the dust intercepts essentially the same amount of sunlight in all colors. The reddish color of the sky is produced when the blue light is absorbed by the dust, but the red light is scattered throughout the sky.

It is not clear if Martian hazes are due to dust or ice. They have optical depths on the order of 1, although they can sometimes can be greater than one [Christensen and Zurek, 1984]. Viking 1 found a background haze of $\tau \sim 0.3$, while Viking 2 found a similar haze of $\tau \sim 0.5$. The optical depth for Martian dust aerosols is 0.5 as listed in Table 3-4. As a comparison, Martian dust storms have much higher optical depths. Viking Lander 1 measured higher optical depths of $\tau = 2.7$ and 3.7 for two global storms [Tillman et al., 1979]. Pollack et al. [1977 and 1979] estimated an upper limit on those dust storms of $\tau = 3.7$ and 9, respectively.

A dust devil is a swirling, vertical updraft of air developed by local heating of the air above a flat desert floor. Pathfinder detected several signatures of a dust devil that passed over the lander on Sol 25 [Schofield et al., 1997]. Over a period of approximately two minutes, the surface pressure showed a sharp minimum approximately 0.5% below the background pressure. The East wind increased suddenly as the dust devil approached the lander, and the pressure began to fall. As the dust devil passed over the lander, pressure began to rise, the East wind died away, and the West wind increased suddenly. Finally, as the dust devil moved away, pressure returned to normal, and the West wind died away. This is a textbook dust-devil signature. The motion direction of the dust devil relative to the Pathfinder in terms of wind speed and surface pressure is schematically shown in Figure 3-10.

The Pathfinder's Rover measured the dust deposited on the Rover's solar array by measuring the change in transparency of a movable glass cover as dust settled on it [Rieder et al., 1997; Team, 1997]. The Rover solar array was found to be accumulating dust at a rate of about a quarter of a percent of coverage per day. This is very close to the coverage of 0.22% predicted [Landis, 1996]. The deposition rate seems to be the same on the sols when the Rover was in motion as it was on sols when the Rover remained in place, indicating that the deposition was probably due to dust settling out of the atmosphere, not dust kicked up by the rover's motion.

Mars Pathfinder produced the following findings about dust aerosols [Schofield et al., 1997; Team, 1997]:

1. Martian dust included magnetic composite particles with a mean size of one micron.

2. The observed atmospheric clarity was higher than was expected from Earth-based microwave measurements and Hubble Space Telescope observations.

3. Dust was confirmed as the dominant absorber of solar radiation in the Martian atmosphere, which has important consequences for the transport and circulation of energy in the atmosphere.

4. Frequent "dust devils" were found with an unmistakable temperature, wind, and pressure signature, and with morning turbulence; at least one may have contained dust (on Sol 62), suggesting that these gusts are a mechanism for mixing dust into the atmosphere.

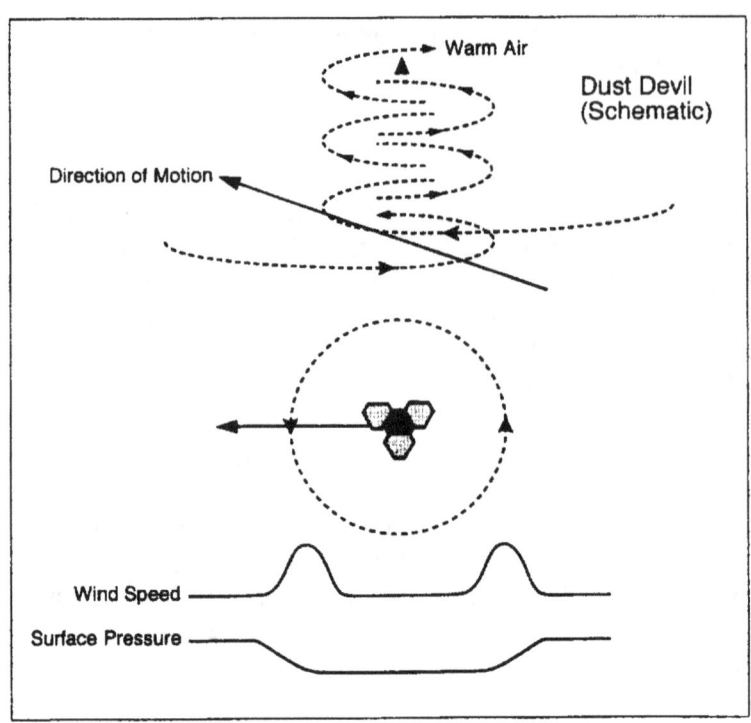

Figure 3-10. Simplified Schematic Drawing of the Dust Devil That Passed Over the Sagan Memorial Station (Mars Pathfinder) on Sol 25. It shows direction of motion and the graphs of a textbook dust devil in terms of wind speed and surface pressure.

It is expected that total attenuation at Ka-band due to Martian dust aerosols is less than 0.1 dB along a vertical path.

3.5 Communication Blackout During Atmospheric Entry Phase

When a high-speed (supersonic) spacecraft enters the Martian atmosphere, because its flight speed is much greater than the local speed of sound, a shock layer is formed in the front of the capsule body. The plasma results from thermal ionization of the constituents of the air as it is compressed and heated by the strong bow shock or heated within the boundary layer next to the surface. Because the plasma density surrounding the capsule is very high, communications are disrupted during the entry phase. This phenomenon is usually known as a blackout. In the 1960s during the Apollo Moon mission period, NASA conducted much research and many experiments for Earth atmospheric reentry [Tischer, 1963; Cuddihy et al., 1963; Huber, 1965; Olte, 1966; Mitchell, 1967; Dunn, 1970]. For the manned Apollo mission, the interaction of the high-speed capsule with the dense Earth atmosphere caused a communications blackout duration of 4 to 10 minutes at X-band. During this period, it was not known whether the capsule was destroyed or had just lost contact. Recent literature reflects the lack of ongoing research. Actually, this type of problem has never been completely analyzed [Chadwick, 1997].

The main cause of blackout is reflection or absorption of electromagnetic wave energy at all communication frequencies (f) lower than the local plasma frequency (f_p), where f_p (MHz) = 9.0 $\times 10^{-3}$ $N^{1/2}$ (cm^{-3}). For $f < f_p$ a plasma behaves like a conductor, while for $f > f_p$ the plasma is practically transparent. The critical plasma densities for various frequencies of signals from UHF to Ka-band are listed in Table 3-5.

Table 3-5. Critical Plasma Densities and Communication Frequencies

Signal Frequency	UHF 381 MHz	S-band 2.295 GHz	X-band 8.43 GHz	Ka-band 32 GHz
Plasma Density	1.8×10^9 cm^{-3}	6.5×10^{10} cm^{-3}	8.8×10^{11} cm^{-3}	1.27×10^{13} cm^{-3}

Although the Martian atmosphere is very tenuous compared to Earth's atmosphere, the heat generated at the hypersonic entry velocity of 7.0 km/s due to the atmospheric drag can be substantial, and subsequent ionization of some of the atmospheric gases may occur. Therefore, a shock-induced envelope of ionized gases (i.e., a plasmasheath) may form around and trail the capsule as it descends through the Martian atmosphere.

There is no simple solution for the plasma density distribution around the capsule because this process involves complicated chemical reactions. The production rate of electrons due to the impact is a function of spacecraft speed, capsule shape, atmospheric density, and atmospheric composition. Thus, the severity of the attenuation will depend largely on the capsule entry trajectory and altitude profiles of velocity, atmospheric density, and composition. Behind strong normal shock waves, the predominant electron production processes are multi-body atom-atom ionizing collision, photoionization, electron impact, atom-molecule collisions, and molecule-molecule collisions [Lin and Teare, 1963]. The resultant electron concentration also depends on entry angle and attack angle. Vertical incidence (90°) usually generates a great deal of ionization, while a small entry angle (near horizontal motion) causes weaker ionization.

In general, there are two distinctly different plasma regions around a capsule during its impact ionization period, as shown in Figure 3-11. The outer plasmasheath region (between the shock wave and the free shear boundary layer) consists of inviscid air, which is compressed and heated by a strong bow shock wave, producing free electrons. When this fluid expands around the capsule, some of the electrons and ions recombine, but the flow is never in complete thermodynamic and chemical equilibrium [Huber, 1965; Grose et al., 1975]. The inner plasma region (between the free shear layer and the rear body surfaces) is one of separated fluid, which is composed of boundary layer air "contaminated" with ablation material from the heat shield. This air mixture recirculates in the base flow region and is finally scavenged into the wake. There is a higher electron density in the stagnation region (front nose) where electrons are produced, while there is a low plasma density in the rear wake region of the capsule where the antenna is installed. Thus, the signal propagation from an antenna on the body will be through both the inviscid and the separated regions.

To obtain a numerical solution of plasma density profiles around a capsule, one usually needs to solve the Navier-Stokes equations for non-viscous, or viscous flow with gases in thermochemical equilibrium and nonequilibrium [Boukhobza, 1997; Dunn, 1970, Huber, 1965]. Plasma density calculations also need to include all ion chemical reaction (production and recombination) equations of charge exchanges for all species. Some simplified calculation methods to roughly estimate the plasma effects have been worked out [Tischer, 1962 and 1963].

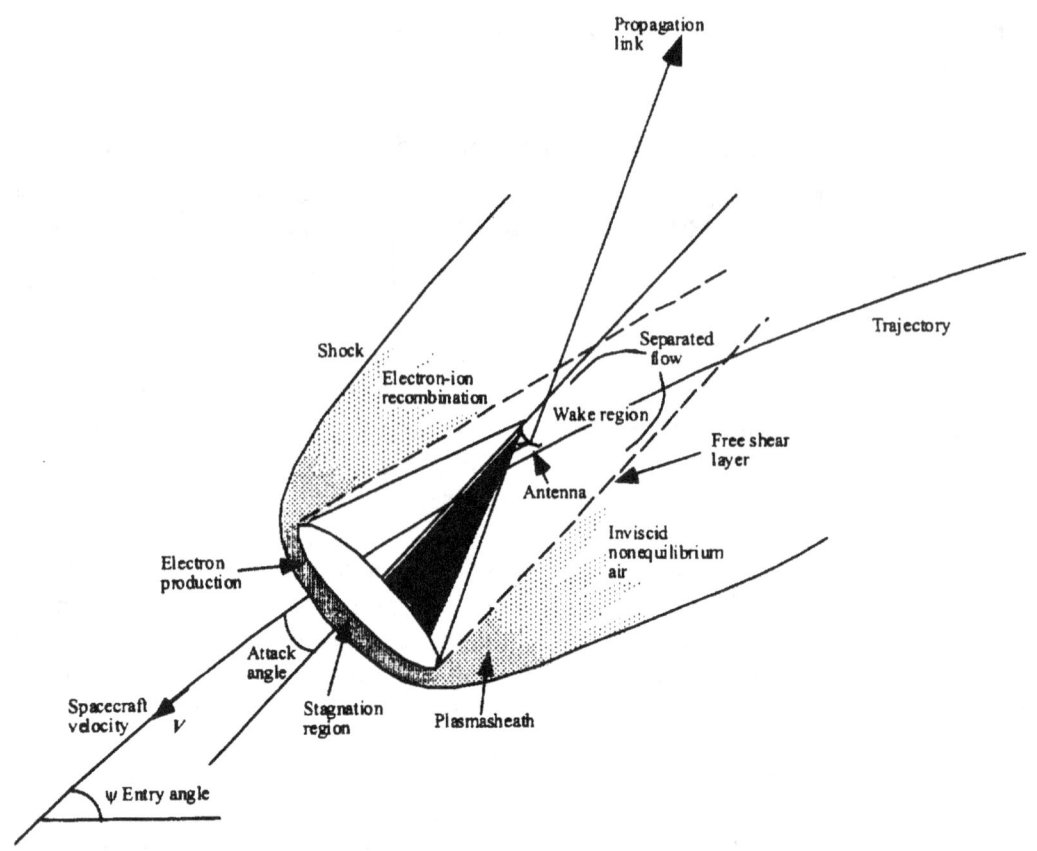

Figure 3-11. Diagrammed View of a Blunt Hypersonic Spacecraft Entering the Martian Atmosphere. A plasmasheath generated around the capsule blacks out the communication signals.

To study the blackout problem caused by atmospheric impacting ionization, JPL developed a computer program to calculate the equilibrium electron density through the thermochemistry and shock heating [Horton, 1964]. The program initializes with a specific gas mixture, temperature, and pressure. For a particular shock velocity, various ion chemical compositions are generated under an equilibrium thermodynamic state. Equations (3-12) and (3-13) below, respectively, represent an empirical fit for the resultant electron density for a shock velocity between 4.0 and 9.0 km/s, at a typical interest range, at stagnation and at the wake region of the capsule.

$$n_{e,s} = 1.5 \times 10^{10} \, \rho^{0.95} V^{11.8} \quad \text{at stagnation point} \tag{3-12}$$

$$n_{e,s} = 1.83 \times 10^{9} \, \rho^{0.95} V^{10.37} \quad \text{at wake region} \tag{3-13}$$

where ρ is neutral atmospheric mass density in g/cm^3, while V is velocity of the capsule in km/s. Using this computer program and an isothermal Martian atmospheric density profile, Spencer [1964] found the electron density in the wake with a peak value of about 3×10^{12} cm^{-3}. This density is higher than critical density for X-band communication, as shown in Figure 3-12. Using a different Martian atmospheric model, calculations by Nordgard [1976] showed that the electron

38

density in the wake region of the capsule had a peak value of 7.9×10^{10} cm^{-3}. This is above the critical electron density for the S-band communication links, but it will not black out the X-band transmission link.

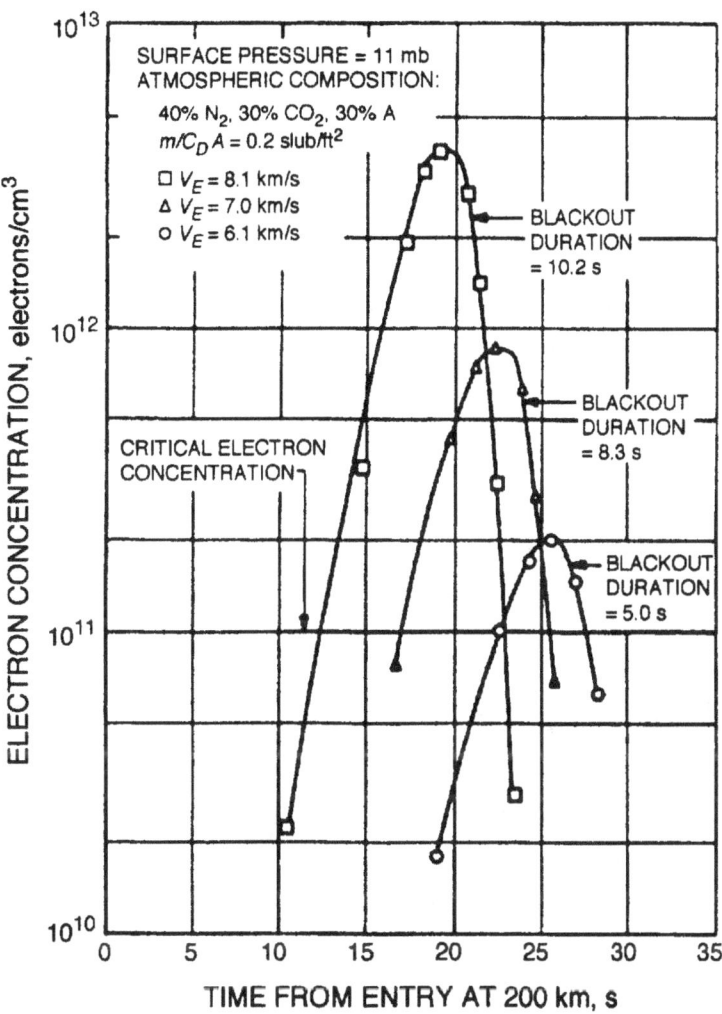

Figure 3-12. Electron Densities in the Capsule Wake Region versus Time from Entry for Various Entry Velocities and for Entry Angle ψ = 90° (from Spencer, 1964).

The Viking I and II landers, the Mars Pathfinder, and future missions (such as Mars 2003) all involve various Martian atmospheric entry projects. The former Soviet Union's Mars 6 experienced a 61-s signal-loss interval at 142.3 MHz, starting from 75-km altitude [Vasilleva et al., 1975]. During the Viking mission, the two Viking landers experienced 1-minute communication blackouts at the UHF band during their EDL phase. On July 4, 1997, the Mars Pathfinder (MPF) spacecraft entered the atmosphere of Mars after a 7-month cruise. The entry speed was approximately 7 km/s. The telecommunication frequency was X-band (8.43 GHz). During descent, signals transmitted from the MPF were extremely dynamic [Wood et al., 1997]. At 125 km in altitude, the MPF speed was 7.6 km/s. The spacecraft used a backshell low-gain antenna to communicate with Earth. During the peak deceleration period, 220 s before the landing, there was a 30-s signal outage as shown in Figure 3-13. This 30-s communication gap was possibly due to blackout. It may also be due to some other effects. This suggests a much higher electron density around the capsule than expected by the model in previous studies (as shown in Figure 3-12). Thus, a more accurate plasma distribution should be recalculated using an updated Martian atmospheric density and composition model.

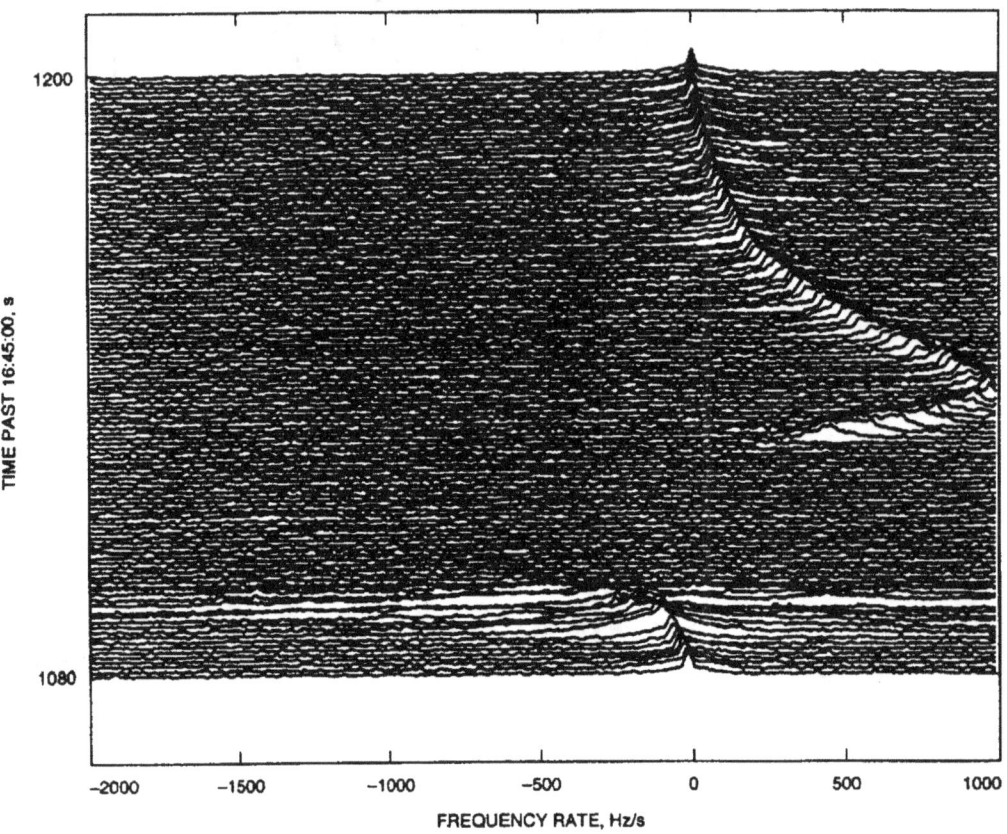

Figure 3-13. Tracking Signals During Mars Pathfinder Atmospheric Entry Phase. The plot shows the Doppler frequency peak ramp rate (i.e., signal derivative) vs. time for the peak deceleration event. There was a 30-s signal outage beginning at 17:03:20 UTC (1100 s past 16:45:00) (from Wood et al., 1997).

Currently, some of the proposed solutions for communication blackout are:

1. Place the antenna where the plasma is diluted (e.g., on the lee side) and communicate by a relay orbiter.

2. Inject some electron-absorbing liquid chemicals into the flowfield to neutralize the plasma [Cuddihy et al., 1963].

3. Increase the frequency of transmission signals from X-band into Ka-band [Brummer, 1963].

4. Apply a magnetic field to the plasma and allow the whistler-mode wave to propagate through a gyrofrequency window [Leblanc and Fujiware, 1996].

For near future Martian atmospheric entry and manned landing programs, the first three methods are more feasible solutions for overcome the communication blackout effects, but finding the best solution remains to be experimentally tested.

3.6   Summary and Recommendations

Compared to Earth's troposphere, the Martian troposphere has a relatively small effect on radio wave propagation. The refractive index of the Martian troposphere is about two orders of magnitude smaller than that of Earth's. Attenuation due to clouds and fog depends largely on their water contents. So far we have little knowledge because direct cloud measurements are not yet available. However, Martian clouds are expected to have relatively less water liquid content because the clouds have a small optical depth. At most, the clouds are expected to be similar to terrestrial high-level cirrus clouds. Martian fog and aerosols (haze) have also been found to have little optical depth.

Even though the Martian tropospheric radio refractivity has a small value, it can still cause ray bending and multipath effects. We recommend that researchers perform an accurate calculation on the excess phase and group delays (range and time delays). Other effects (such as range rate errors, appearance angle deviation, defocusing loss on Mars, etc.) also need to be estimated. Ice depolarization effects on radio waves are still unknown, although they are expected to be small because of the lower optical depth and thinner cloud layer.

Although not as pronounced as on Earth, clouds are a common feature on Mars. The Martian atmosphere has only a trace of water vapor; however, the temperature and pressure are such that the atmosphere is usually close to saturation and produces clouds. Even from Earth-based telescopes, clouds have been observed by transient brightening on the surface of Mars. Numerous cloud patterns have been seen from the Mariner and Viking spacecraft and have been classified into various categories [Carr, 1981; French et al. 1981]. The optical depths of Martian clouds and fogs are about 1.0. Thus, it is expected that they have little attenuation on microwave propagation and optical communications.

The optical depth of Martian aerosol dust is about 0.5. Its attenuation effect on radio wave propagation is smaller than that of the Martian clouds. However, the amount of Martian aerosols should be monitored, and the maximum height to which aerosols extend in the atmosphere should be determined. Also, the size and shape of the aerosol particles should be measured.

In general, the Martian atmospheric environment is quite good for optical communications because of its thinner atmosphere, except during dust storms because the optical depth of Martian clouds is only one fifth of Earth clouds. Martian aerosols can cause some attenuation to laser beams; however, this effect is very small compared with that of aerosols at Earth.

During spacecraft atmospheric entry, signal transmissions will be significantly degraded (blackout). Even though a great effort has been made in the past 30 years, this problem has never been completely solved. A 30-s communication disruption at X-band during the Mars Pathfinder descent was probably caused by plasmasheath blackout. Two basic directions for reduction of the problem are indicated:

(1) Increase the signal frequency to a point where its value is higher than the plasma frequency value.

(2) Reduce the plasma electron density by modifying the plasma.

References

Annis, J., The atmosphere of Mars and optical communications, *TDA Progress Report 42-91*, Jet Propulsion Laboratory, Pasadena, CA, 124, November 15, 1987.

Battan, L.J., *Radar Observation of the Atmosphere*, U. of Chicago Press, 1973.

Bean, B.R. and E.J. Dutton., *Radio Meteorology*, Dover Publications Inc., New York, USA, 1966.

Boukhobza, M., Communication blackout during earth-atmosphere entry of capsules, *NASA Technical Report*, 97N25663, May, 1997.

Briggs, G., et al., Martian dynamical phenomena during June–November 1976: Viking orbiter results, *J. Geophys. Res., 82*, 4121, 1977.

Brummer, E.A., X-band telemetry systems for reentry research, *NASA Technical Report* 63N14953, April, 1963.

Carr, M.H., *The surface of Mars*, Yale University Press, 1981.

Chadwick, K.M., Plasma and aerothermal measurements on a hypervelocity re-entry vehicle, *J. Spacecraft Rockets, Vol. 34*, 602-608, 1997.

Christensen, P.R. and R.W. Zurek, Martian north polar hazes and surface ice: Results from the Viking survey/composition mission, *J. Geophys. Res., 89*, 4587, 1984.

Cuddihy, W.F., et al., A solution to the problem of communications blackout of hypersonic reentry vehicles, *NASA Technical Report*, 71N70235, Oct. 1963.

Dunn, M.G., Comparison between predicted and measured blackout boundaries for earth of Apollo, *NASA Technical Reports*, 71N21104, Jan. 1970.

French, R.G., et al., Global patterns in cloud forms on Mars, *Icarus, 45*, 468, 1981.

Golombek, M.P., et al., Overview of the Mars Pathfinder mission and assessment of landing site predictions, *Science, 278*, 1743, 1997.

Grose, W.L., et al., An analysis of communications blackout for Pioneer Venus entry probes, *AIAA Paper 75-182*, Jan. 1975.

Gunn, K.L.S., and T.W.R. East, The microwave properties of precipitation particles, *Quart. J. Roy. Meteor. Soc., vol. 80*, 522, 1954.

Hinson, D.P., F.M. Flasar, R.A. Simpson, J.D. Twicken, and G.L. Tyler, Initial Results from Radio Occultation Measurements with Mars Global Surveyor, *J. Geophy. Res.-Planet, 104*, 26297, 1999.

Horton, T.E., The JPL thermochemistry and normal shock computer program, Jet Propulsion Laboratory, Pasadena, CA, *Technical Report No. 32-660*, Nov. 1, 1964.

Huber, P.W., Research approach to reentry communications blackout, Presented in the Conference on Langley research related to Apollo Missions, *NASA SP-101*, June, 22, 1965.

Kahn, R., The spatial and seasonal distribution of Martian clouds and some meteorological implications, *J. Geophys. Res., 89*, 6671, 1984.

Keating, G.M., et al., The structure of the upper atmosphere of Mars: In situ accelerometer measurements from Mars Global Surveyor, *Science, 279*, 1672, 1998.

Landis, G., Dust obscuration of solar arrays, *Acta Astronautica, Vol. 38*, No. 11, 895-891, 1996.

Leblanc, J.E., and T. Fujiware, Comprehensive analysis of communication with a reentry vehicle during blackout phase, *Trans. Japan Soc. Aero. Space Sci., Vol. 39*, 211-221, Aug. 1996.

Leovy, C., Martian meteorological variability, The Mars reference atmosphere, Ed. by Kliore, A.J., *Adv. Space Res., 2*, 2, 1982.

Lin, Shao-Chi and J.D. Teare, Rate of ionization behind shock waves in air. II. Theoretical interpretations, *Phys. Fluids, Vol. 6*, 355-375, Mar. 1963.

Ludlam, F.H., *Clouds and Storms*, Pennsylvania State U. Press, 1980.

Mitchell, F.H., Jr., Communication-system blackout during reentry of large vehicles, *Proceedings of the IEEE, Vol. 55*, 619-626, 1967.

Nordgard, J.D., A Martian entry propagation study, *Radio Sci., Vol., 11*, 947-957, 1976.

Olte, A., Study of plasma sheath associated with communications blackout: Final report, *NASA Technical Report 67N19139*, Dec. 1966.

Pollack, J.B. et al., Properties of aerosols in the Martian atmosphere, as inferred from Viking Lander imaging data, *J. Geophys. Res., 82*, 4479, 1977.

Pollack, J.B. et al., Properties and effects of dust particles suspended in the Martian atmosphere, *J. Geophys. Res., 84*, 2929, 1979.

Rieder, R. et al., The chemical composition of Martian soil and rocks returned by the mobile alpha proton X-ray spectrometer: Preliminary results from the X-ray mode, *Science, 278*, 1771, 1997.

Schofield, J.T. et al., The Mars Pathfinder atmospheric structure investigation/meteorology (ASI/MET) experiment, *Science, 278*, 1752, 1997.

Seiff, A., Post-Viking models for the structure of the summer atmosphere of Mars, The Mars reference atmosphere, Ed. by Kliore, A.J., *Adv. Space Res., 2,* 2, 1982.

Slobin, S.D., Microwave noise temperature and attenuation of clouds: Statistics of these effects at various sites in the United States, Alaska, and Hawaii, *Radio Sci.,* 17, 1443, 1982.

Spencer, D.F., An evaluation of the communication blackout problem for a blunt Mars-entry capsule and a potential method for the elimination of blackout, Jet Propulsion Laboratory, Pasadena, CA, *Technical Report No. 32-594,* April 15, 1964.

Team, R., Characterization of the Martian surface deposits by the Mars Pathfinder rover, Sojourner, *Science, 278,* 1765, 1997.

Thorpe, T., Viking Orbiter observations of atmospheric opacity during July–November 1976, *J. Geophys. Res., 82,* 4151, 1977.

Tillman, J.E. et al., Frontal systems during passage of the Martian north polar hood over the Viking Lander 2 site prior to the first 1977 dust storm, *J. Geophys. Res., 84,* 2947, 1979.

Tischer, F.J., A rough estimate of the "blackout" time in re-entry communications, *NASA Technical Report 64N29516,* GSFC, June, 1962.

Tischer, F.J., Communication blackout at re-entry, *NASA Technical Report 64N13335,* Sept. 1963.

Vasilleva, L.I., et al., Determination of the duration and limits of communication loss upon entry of the descent vehicle of the space probe Mars 6 into the Martian atmosphere, *Kosmich. Issled., 14,* 610, 1976.

Wood, G.E., et al., Mars Pathfinder entry, descent, and landing communications, *TDA Progress Report 42-131,* Jet Propulsion Laboratory, Pasadena, CA, 1–19, November 15, 1997.

4. Martian Atmospheric Gaseous Attenuation

4.1 Introduction

When radio waves pass through an atmosphere, the waves suffer molecular absorption and scattering at centimeter and millimeter wavelengths [CCIR, 1986; Crane, 1981; Liebe, 1981; Waters, 1976]. At Earth, the gaseous absorption is due primarily to atmospheric water vapor and oxygen. There are 29 absorption lines for H_2O (up through 1097 GHz) and 44 lines for O_2 (up through 834 GHz). Relatively narrow and weaker ozone (O_3) lines are above 100 GHz. Between 120 GHz and 1097 GHz, water vapor plays a serious role in the radio wave attenuation. For frequencies greater than 70 GHz, other gases can also contribute an attenuation in the absence of water vapor; however, their spectral lines are usually too weak to affect propagation [Waters, 1976; Ulaby et al., 1981; Smith, 1982; Liebe, 1985].

The principal interaction mechanism between radio waves and gaseous constituents is molecular absorption from molecules. Accurate predictions of atmospheric attenuation can be determined from radiative transfer calculations. Absorption attenuation of radio waves results from a quantum level change in the rotational energy of molecules. Spectral line absorption occurs when a quantized system, such as a molecule, interacts with an electromagnetic radiation field and makes a transition between two quantum states of the system. The resonant frequency f_{lm} is [Waters, 1976; Ulaby et al., 1981]

$$f_{lm} = (E_l - E_m)/h \qquad (4\text{-}1)$$

where E_l and E_m are energy levels of final and initial rotational energy state, and h is Plank's constant.

The general expression for the absorption coefficient $\kappa(f, f_{lm})$ may be written as

$$\kappa(f, f_{lm}) = \frac{8\pi^3 n_i f \mu^2}{3hcQ} \left\{ e^{-E_l/kT} - e^{-E_m/kT} \right\} g_l \mid \phi_{lm} \mid^2 L(f, f_{lm}) \qquad (4\text{-}2)$$

where n_i is the number of absorbing molecules per unit volume for *ith* species, μ the total dipole moment, g_l the statistical weight of the lower state, ϕ_{lm} the transition matrix element, $L(f, f_{lm})$ a function describing the line shape, and Q the partition function.

The volume absorption coefficient κ describes the interaction of radiation with the absorbing matter. κ is a function of the density of the absorbing substance, the atmospheric temperature, and the pressure. The coefficient has units of cm^{-1}. There is a relation of 1 cm^{-1} = $10^6 \log_{10} e$ dB/km = 4.34×10^5 dB/km. Optical depth is an integration of the absorption coefficient κ along the path, which is dimensionless and can be expressed in a unit "neper" (logarithms to base e) or dB (logarithms to base 10). 1 Np = 4.34 dB.

Absorption of electromagnetic energy by gaseous molecules usually involves the interaction of electric or magnetic field-of-incident waves with an electric or magnetic dipole. H_2O and O_2 are the two major atmospheric constituents for radio wave absorption in the microwave band. The oxygen molecule has a permanent magnetic moment arising from two unpaired electron spins. Magnetic interaction produces a family of rotation lines around 60 GHz and an isolated line at 118.8 GHz. Water vapor is a molecule with an electric dipole. Through an electric interaction with the incident electric field, Water vapor produces rotational lines at 22.2, 183.3, and 323.8 GHz and at several frequencies in the far-infrared band. Each of the absorption spectral lines has a certain width because the energy levels vary when molecules are in motion. Among the various factors causing line broadening, atmospheric pressure broadening is the most important in the microwave band.

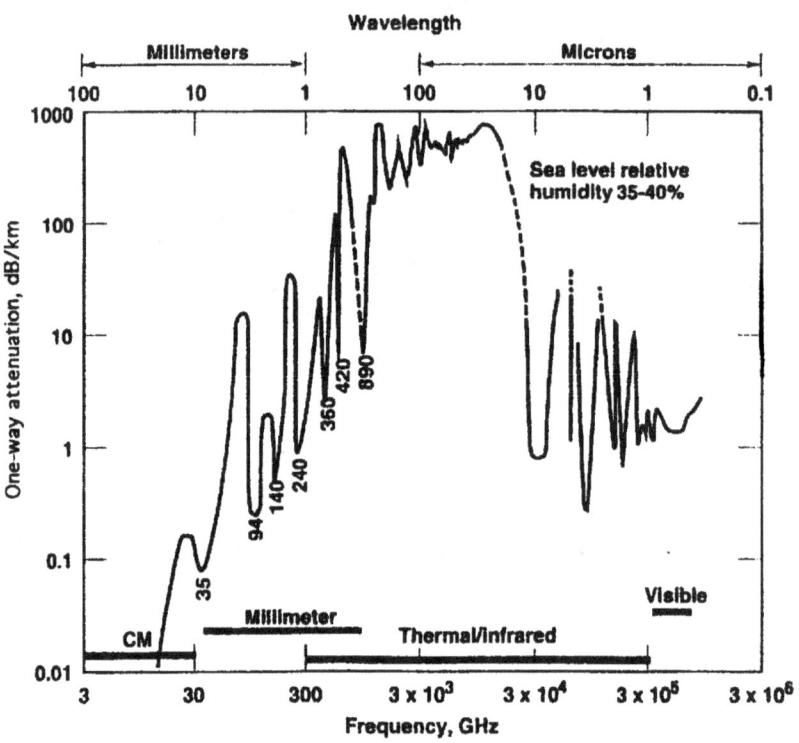

Figure 4-1. Specific Gaseous Attenuation for a One-Way Horizontal Earth's Atmospheric Path in the Frequency Ranges from Microwave to Visible Light.

Figure 4-1 shows the atmospheric attenuation at Earth surface as a function of frequencies (3 GHz to 3×10^6 GHz) in a range from microwave to visible light (wavelength from 100 mm to 0.1 μm). Below 890 GHz, there are seven attenuation peaks, which are caused by H_2O and O_2. Between these absorption maxima, a number of low attenuation atmospheric windows exist for radio wave propagation. The center frequencies of the windows have been marked under each trough. The broad peaks between 890 GHz and 10,000 GHz are mainly due to water vapor. Above 10,000 GHz ($\lambda > 30$ μm) from infrared to the optical range, there are some absorption lines due to CO_2 and N_2.

Argon is a heavy, chemically inert gas. Both CO_2 and N_2 are also very stable gases. Because these molecules do not have an intrinsic electric or magnetic dipole, they do not absorb electromagnetic energy in the microwave frequency range. However, through collisions among themselves when the neutral density is reasonably high, CO_2 and N_2 can generate dipoles and interact with radio waves. Other constituents (such as O_3, SO_2, and N_2O) can also produce absorption lines. However, because their concentration is so small their contribution can be neglected in comparison with effects of H_2O and O_2.

4.2 Martian Gaseous Composition and Comparison With Earth Atmosphere

Attenuation from atmospheric molecules is heavily dependent on atmospheric structure, including its temperature, pressure, composition, abundance, etc. Because the gaseous attenuation for the Earth's atmosphere has been well studied and documented, we should compare the Martian atmospheric structure with Earth's. This comparison can help us understand Martian gaseous attenuation.

The atmosphere of Mars is quite different from that of Earth in composition, abundance, and altitude profiles [Hanson et al., 1977; Owen, 1992; Nier and McElroy, 1977; McElroy et al., 1977]. It is composed primarily of carbon dioxide, with small amounts of other gases. Some basic parameters are:

Surface Pressure: ~6.1 mb (variable)

Surface Density: ~0.020 kg/m^3

Scale height: ~11.1 km

Average temperature: ~210 K

Diurnal temperature range: 184 K to 242 K

Mean molecular weight: 43.34 g/mole

Atmospheric composition (by volume):

Major:

carbon dioxide (CO_2) — 95.32%

nitrogen (N_2) — 2.7%

argon (Ar) — 1.6%

oxygen (O_2) — 0.13%

carbon Monoxide (CO) — 0.08%

Minor (units in parts per million [ppm]):

water vapor (H_2O) — ~100–400 (variable)

nitrogen Oxide (NO) — 100

neon (Ne) — 2.5

hydrogen-deuterium-oxygen (HDO) — 0.85

krypton (Kr) — 0.3

xenon (Xe) — 0.08

ozone (O_3) — 0.04–0.2.

In Figure 4-2a, b, c, and d, we show altitude profiles of the Martian atmospheric density for all compositions in various altitude ranges [McElroy and McConnell, 1971a and b; Anderson, 1974; Chen et al., 1978; Fox and Dalgarno, 1979; Yung et al., 1977]. Most gases decrease in density with increasing altitude, although some light gases (such as atoms of O, H, N) have peak concentrations at higher altitudes.

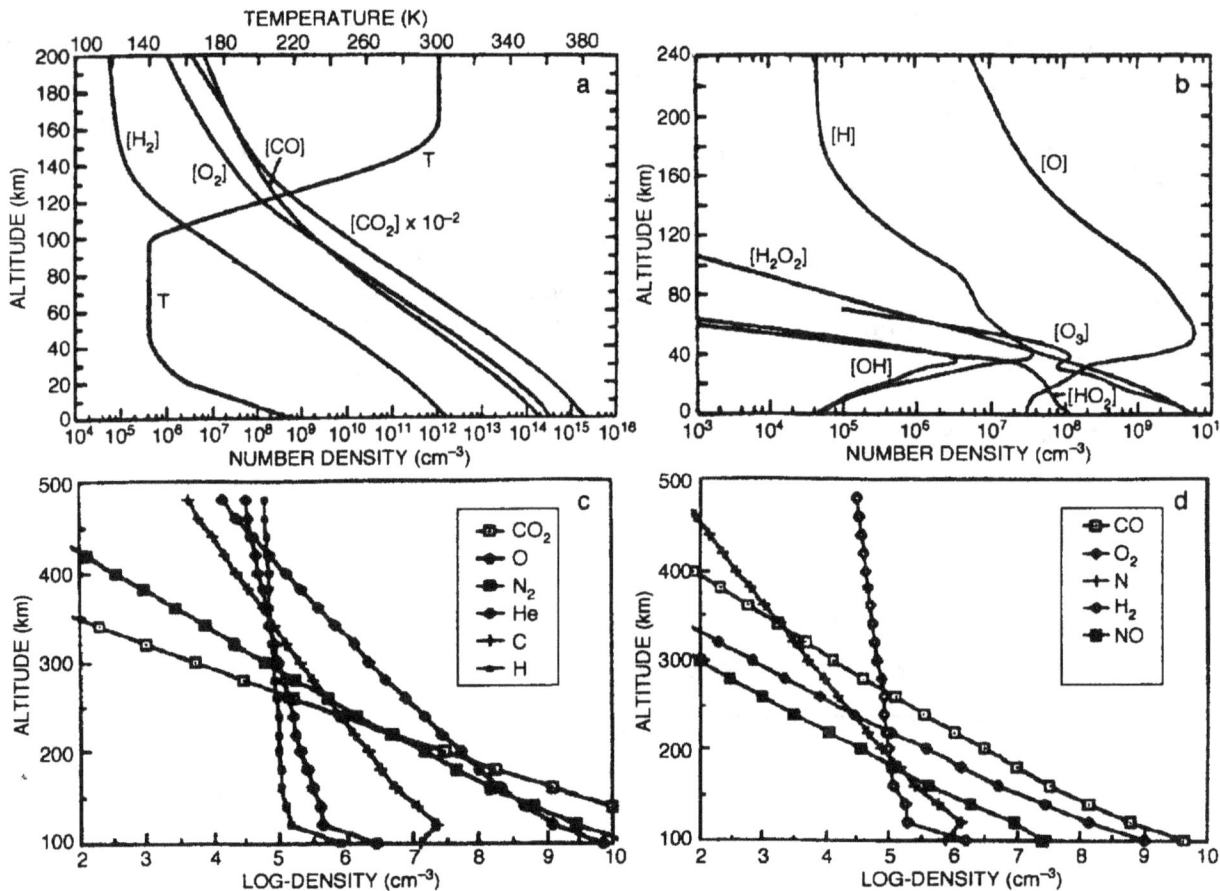

Figure 4-2. Martian Atmospheric Density Profiles for Various Constituents. a) CO_2, CO, O_2, and H_2 in an altitude range between 0 and 200 km, b) O, H, OH, H_2O_2, and O_3 from 0 to 240 km, c) CO_2, O, N_2, He, C, and H between 100 and 500 km and d) CO, O_2, N, H_2, and NO between 100 and 500 km.

As a comparison, we also list parameters of the Earth atmosphere below [NOAA, 1976; Jursa, 1985]. The Earth's atmosphere consists of N_2, O_2, and several minor gases. Their composition profiles are shown in Figure 4-3 [Junge, 1963].

Surface Pressure: 1013 mb (average)

Surface Density: ~1.29 kg/m^3

Scale height: ~9.5 km

Average temperature: ~300 K

Diurnal temperature range: 210 K to 320 K

Mean molecular weight: 28.61 g/mole

Atmospheric composition (by volume):

    Major:

        nitrogen (N_2) — 78.09%

        oxygen (O_2) — 20.95%

        argon (Ar) — 0.93%

        carbon Dioxide (C O_2) — 0.03%

    Minor (units in ppm):

        water vapor (H_2O) —~40–40,000 (variable)

        neon (Ne) — 20

        helium (He) — 5.2

        methane (CH_4) — 1.5

        krypton (Kr) — 1.1

        hydrogen (H_2) — 1.0

        nitrous oxide (N_2O) — 0.6

        carbon Monoxide (CO) — 0.2

        ozone (O_3) — < 0.05

        xenon (Xe) — 0.09

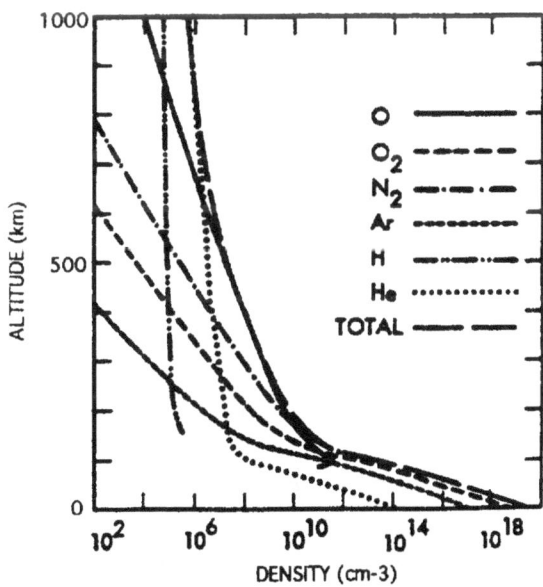

Figure 4-3. Earth Atmospheric Number Density Profiles for Individual Species. Above 120 km in altitude, scale heights are different for each species.

From Figures 4-2 and 4-3, we can note that each constituent has a different scale height above the homopause. For example, at Mars, the scale height for CO_2 is ~8 km, for oxygen (O_2) is

~9 km, and water vapor has a variable scale height. Average scale height for all gases below the homopause is 11.1 km. At Earth, below 120 km, all constituents have similar scale heights (N_2 ~8.7 km; O_2 ~9.0 km; Ar ~9.2 km). Above 120 km (in the thermosphere), scale heights almost double for these gases. The atomic gases O and H occur as larger concentrations at 100 km altitude. In this study, because we are interested in the surface atmospheric attenuation, we only consider gaseous density and scale heights near the surface of planets.

To calculate Martian atmospheric attenuation, we need various atmospheric parameters. In Table 4-1, we have listed all surface average values for both Mars and Earth.

From Table 4-1, we can see that the Martian surface atmospheric pressure is only about 6/1000 of Earth's pressure. The Martian atmospheric average molar weight (gram/mole) is larger than at Earth's because the dominant gas, CO_2, has a larger mass than N_2 at Earth. However, the Martian atmospheric mole volume is much larger (by three orders) than Earth's. Because of low pressure at Mars, the mass density of the Mars atmosphere is 61 times less than that of Earth. Thus, the average number density is also smaller by about 2 orders of magnitude. The scale heights at both planets are only slightly different.

Table 4-1. Surface Atmospheric Parameters at Mars and Earth

Planets	P, pressure (mb)	T, temperature (K)	M, mean molecule weight	ρ, mass density (kg/m³)	N, number density (m⁻³)	V_m, mole volume (m³/kmole)	H, scale height (km)
Mars	6.1	210	43.34 g/mole	0.021	2.85×10^{23}	2.1×10^3	~11.1
Earth	1013	300	28.61 g/mole	1.29	2.7×10^{25}	22	~9.5

In order to make a more accurate parameter comparison, we have defined:

F_i: fraction by volume for ith species (air mix ratio)

β_i: fraction by mass for ith species

p_i: pressure for ith species; P: pressure for all species (mb)

n_i: number density ith species (cm⁻³); N: sum for all species

ρ_i: mass density ith species; ρ: mass density for all species (gram/m³)

M_i: molar weight (i.e., gram/mole) for ith species; M: mean molecule weight for all species (gram/mole)

V_m: air volume for one kilomole at surface (m³/kmole)

N_A: Avogadro constant; k_B Boltzmann constant; R_A: Universal gas constant

T: temperature (K); H: scale height (km), which is defined as $N=N_0e^{-z/H}$

On the basis of gas dynamic theory, the following fundamental relations exist between these parameters in a multi-species atmosphere:

$$p_i = n_i k_B T \ , \ P = N k_B T \ , \ P = \sum_i p_i \tag{4-3}$$

$$F_i = \frac{p_i}{P} = \frac{n_i}{N} \ , \ N = \sum_i n_i \tag{4-4}$$

$$\beta_i = \frac{\rho_i}{\rho} = \frac{F_i M_i}{M} \ , \ M = \sum_i F_i M_i \ , \tag{4-5}$$

$$\rho_i = \frac{n_i M_i}{N_A} = \frac{F_i M_i N}{N_A} = \frac{F_i M_i \rho}{M} = \beta_i \rho \ , \ \rho = \frac{NM}{N_A} \ , \ \rho = \sum_i \rho_i \tag{4-6}$$

$$n_i = \frac{\rho_i}{k_B T} = \frac{F_i \rho N_A}{M} = \frac{\rho_i N_A}{M_i} \ , \tag{4-7}$$

$$V_m = \frac{N_A}{\sum_i n_i} \tag{4-8}$$

Starting from parameters F_i, M_i, and ρ, we have derived parameters, M, β_i, ρ_i, and n_i. These parameters for six main compositions are listed in Table 4-2. In the table, for each species, the second to sixth columns are, respectively, molecular weight (gram molecule), M_i; atmospheric mixing ratio by volume, F_I; fraction by weight, β_i; gaseous mass density, ρ_i; and number density, n_i.

Table 4-2. A Comparison of Atmospheric Compositions Near the Surfaces of Mars and Earth

Gaseous Composition		Mars Surface (6.1 mb, 210 K)				Earth Surface (1013 mb, 300 K)			
mole-cule	M_i, Weight (g/mole)	F_i, mix ratio (by volume)	β_i, fraction in mass	ρ_i, mass density (g/m^3)	n_i, number density (cm^{-3})	F_i, mix ratio (by volume)	β_i, fraction in mass	ρ_i, mass density (g/m^3)	n_i, number density (cm^{-3})
CO_2	44.02	95.32%	96.77%	20.32	2.8×10^{17}	400ppm	615ppm	0.8	1.1×10^{16}
N_2	28.02	2.7%	1.74%	0.365	7.8×10^{15}	78.09%	76.5%	986.9	2.1×10^{19}
Ar	39.96	1.6%	1.48%	0.311	4.7×10^{15}	0.93%	1.3%	16.8	2.6×10^{17}
O_2	30.00	0.13%	900ppm	0.02	3.8×10^{14}	20.95%	21.97%	283.7	5.7×10^{18}
CO	28.00	800ppm	517ppm	0.011	2.3×10^{14}	0.2 ppm	0.2ppm	2.6×10^{-4}	5.6×10^{12}
H_2O	18.02	300ppm	125ppm	0.0026	8.8×10^{13}	1.0%	0.63%	8.1	2.7×10^{17}

* ppm = part(s) per million.

From Table 4-2 we can see that except for CO_2 and CO, all gaseous densities at Mars are less than that at Earth. Even though CO_2 is the dominant gas at Mars, its density is only 25 times more than at Earth. Water vapor abundance is much more variable at both planets. At Mars water vapor abundance ranges from 100 to 400 ppm with an average of 300 ppm, varying with season and latitude as shown in Figure 4-4 [Doms, 1982; Farmer and Doms, 1979]. The dominant feature was found to be the large amount of water vapor over the residual northern cap in the northern summer. At Earth the water vapor abundance is from 40 to 40,000 ppm with an average value of 10,000 ppm.

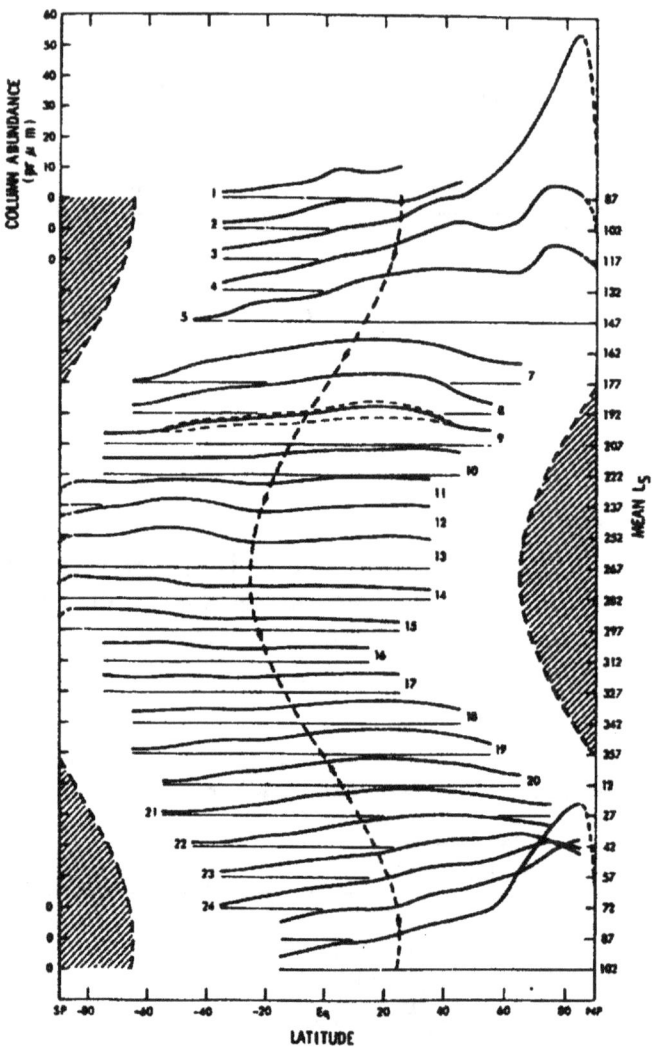

Figure 4-4. Distribution of Water Vapor in the Martian Atmosphere by Latitude and Season (Planetocentric Longitude). The scale of water vapor abundance is shown in the left top. The dashed line through the center of the figure indicates the latitude of the subsolar point. The main feature is the large amount of water vapor over the residual northern cap in the northern summer (from Farmer and Doms, 1979).

Finally, as a comparison, Table 4-3 lists various ratios of atmospheric compositions between Earth and Mars. These ratios are important scale factors related to atmospheric attenuation at both planets.

Ratios for β_i for all species are 1.51 times greater than ratios for F_i because Earth's mean molecular weight is less than Mars' by 1.51. Ratios of Earth relative to Mars for ρ_i (gaseous mass density) are the same as for n_i (number density) because ρ is proportional to n. There is much more N_2, O_2, and H_2O at Earth than at Mars (by factors of 10^3–10^4 in density). We have more interest in the ratios of ρ_i and n_i, in oxygen and water vapor between Earth and Mars because both gases have strong absorption on Earth. At Mars the amounts of O_2 and H_2O are very small, and the question is how much smaller the atmospheric attenuation can be compared to Earth.

Table 4-3. Ratios of Atmospheric Constituents between Earth and Mars

Ratios (Earth/Mars)	CO₂	N₂	Ar	O₂	CO	H₂O
for F_i (fraction by volume)	4.2×10^{-4}	28.9	0.58	161	2.4×10^{-4}	33.3
for β_i (fraction by weight)	6.4×10^{-4}	44	0.88	244	3.9×10^{-4}	50.4
for ρ_i and n_i (density)	0.04	2704	54	1.4×10^{4}	0.024	3068

4.3 Martian Atmospheric Absorption Effects on Microwaves

CO_2 and N_2 are the dominant gases in the Martian atmosphere, and these gases have little absorption in the microwave range. Consequently, O_2 and H_2O cause most of the attenuation. There are 10 strong absorption lines for H_2O below 450 GHz and 39 lines for O_2 between 50 and 70 GHz. Below 300 GHz, ozone (O_3) has 65 weak and narrow absorption lines. There are more than a thousand spectral absorption lines in the microwave and infrared frequency bands for all species. Precise measurements in frequencies and intensities for these lines have been well documented in both the Earth atmosphere and the laboratory [Liebe, 1981; U.S. Department of Commerce, 1968; Waters, 1976; Crane, 1981; Gordy and Cook, 1984].

We have calculated the total atmospheric attenuation, using equations and procedures developed by Waters [1976], Liebe [1981], and Ulaby et al. [1981]; using planetary atmospheric parameters listed in Tables 4-1, 4-2, and 4-3; and using standard spectral intensity measured in the laboratory. The total Martian atmospheric absorption coefficient is a combination of both H_2O and O_2.

$$\kappa_a(f) = \kappa_{H_2O}(f) + \kappa_{O_2}(f), \text{ dB/km} \tag{4-9}$$

Figure 4-5 shows the calculated specific atmospheric attenuation for a horizontal path at the Earth surface and the Mars surface. The plots only give a microwave frequency range (<350 GHz). Above 350 GHz, absorption spectral lines from all species are so complicated that we do not describe them. The dotted lines are for O_2 absorption only. There are two peaks at 60 and

118.8 GHz. The dashed lines are for water vapor only. The absorption peaks are at 22.2, 183.3, and 323.8 GHz. The solid lines are for both combined.

We found that the attenuation values due to oxygen at Mars are reduced by a factor of 1.4×10^4 relative to Earth. This is the ratio ρ_{O2} (oxygen density ratio) between the Earth's atmosphere and the Mars atmosphere at the surface. The water vapor attenuation at Mars is lower by a factor of 3068 than at Earth. This calculation is also dependent on the types of spectral-line shape function chosen [Van Vleck, 1987; Waters, 1976]. There are several types of line shape functions, such as Lorentzian, van Vleck-Weisskopf, and kinetic-line (Gross) shape. Here we have used a kinetic-line shape. We have also not considered the Zeeman effect of the spectral line of the O_2 molecules because Mars has a very weak magnetic field.

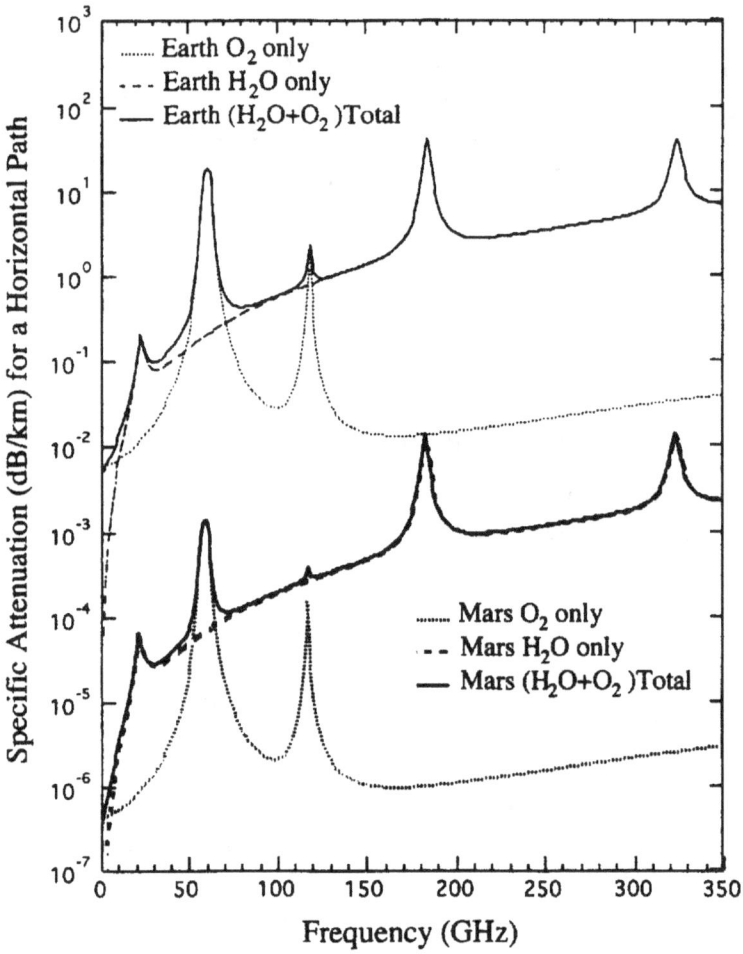

Figure 4-5. Gaseous Specific Absorption Attenuation by Water Vapor, Oxygen, and Both at the Surface of Earth and Mars. The upper three thin lines are for attenuation at Earth, while the lower three thick lines are for Mars.

The optical depth (or opacity) along the zenith, τ_0, is defined as an integration of the absorption coefficient over a vertical path from the surface to infinity:

$$\tau_0 = \int_0^\infty \kappa(z)dz \qquad (4\text{-}10)$$

This is equivalent to integrating all resonant particles along a column because the absorption coefficient is proportional to the particle number density. Actually, the column number for each species is equivalent to a slab with a density of N_0 and a thickness H_0, where H_0 is the scale height of the species, because

$$\int_0^\infty Ndz = \int_0^\infty N_0 e^{-z/H_0} dz = N_0 H_0 \qquad (4\text{-}11)$$

Thus, for each species, the total zenith attenuation (dB) is the product of the surface specific attenuation (dB/km) plotted in Figure 4-5 and the scale height, H_0(km), for an isothermal atmosphere.

At Ka-band (32 GHz), the total zenith attenuation of the Martian atmosphere is estimated to be less than 0.01 dB, assuming that water vapor has a scale height of 10 km. In comparison, in the Earth's atmosphere the attenuation is about 0.3 dB. At higher frequencies, (for example, 100 GHz) the Martian atmospheric attenuation increases to 0.1 dB. Such a small attenuation is negligible for telecommunication.

For an oblique path with zenith angle θ, when $\theta < 70°$, the opacity may be expressed as

$$\tau(\theta) = \tau_0 \sec \theta \qquad (4\text{-}12)$$

Thus, an oblique path will have greater attenuation.

On the other hand, gaseous absorption also can be used as a remote-sensing tool to study Martian atmospheric structures. In the near-infrared and infrared regions there are many absorption lines caused by CO, CO_2, and H_2O. The infrared spectrometer (ISM) on board the Soviet Phobos 2 spacecraft made spectral measurements of the Martian atmosphere [Combes et al., 1991]. There are strong absorption lines at 1.14, 1.84, and 2.55 μm for H_2O; at 2.35 and 5 μm for CO; and at 2.0 and 2.72 μm for CO_2. These results will help us to determine the atmospheric constituents. One application is to use measurements of the column density of CO_2 between the spacecraft and the Martian surface to derive surface pressure and topography maps [Combes et al., 1991].

4.4   Summary and Recommendations

Mars has a much lower atmospheric gaseous attenuation than Earth in the microwave frequency band because of much lower abundances of uncondensed H_2O and O_2 in its atmosphere. Preliminary calculations show that Martian gaseous absorption is at least a factor of 3000 lower than at Earth. For a radio wave with a frequency of 100 GHz through a vertical atmospheric path, the attenuation is estimated to be less than 0.1 dB. However, this calculation is only based on Martian surface composition values. An accurate calculation requires scale heights for each species at different altitudes in the Martian atmosphere. So far we do not have such information. We also do not know how severe the gaseous attenuation can be in the infrared and visible frequency range because of the very complicated spectral lines in these frequencies.

The Martian atmosphere is dominated by CO_2 and N_2. Under normal conditions, these gases do not have electric or magnetic dipoles, so they do not absorb electromagnetic energy from radio waves. However, they may generate dipoles through collisions and interact with waves in the higher frequency range. We often see that both gases having many absorption lines in the infrared and visible ranges in the Earth atmosphere. It should be researched whether or not CO_2 and N_2 at Mars can generate such dipoles.

In this calculation, we have used an average surface value (300 ppm) for Martian water vapor. We have not used a maximum value (400 ppm) that corresponds to the worst case. The error bars can be as large as 33%. Actually, the exact amount of Martian atmospheric water vapor is still controversial, although it is known to depend strongly on latitude and season. An accurate water vapor altitude profile at Mars is not available yet. A conservative estimate for the worst situation of Martian atmospheric attenuation is an increase by a factor of two.

References

Anderson, D.E., Jr., Mariner 6, 7, and 9 ultraviolet spectrometer experiment: Analysis of hydrogen Lyman alpha data, *J. Geophys. Res., 79*, 1513, 1974.

CCIR, Attenuation by atmospheric gases, Report 719-2, 167, in Vol. V, *Propagation in Non-ionized Media, Recommendation and Report of the CCIR*, Geneva: ITU, 1986.

Chen, R.H. et al., The Martian ionosphere in light of the Viking observations, *J. Geophys. Res., 83*, 3871, 1978.

Combes, M., et al., Martian atmosphere studies from the ISM experiment, *Planet. Space Sci., 39*, 189, 1991.

Crane, R.K., Fundamental limitations caused by RF propagation, *Proc. IEEE, 69*, 196, 1981.

Doms, P.E., Water vapor in the Martian atmosphere: A discussion of the Viking data, in *The Mars Reference Atmosphere*, Special Issue in COSPAR, Ed. by Kliore, A.J., *Adv. Space Res., 2*, 1982.

Famer, C.B., and P.E. Doms, Global seasonal variation of water vapor on Mars and the implications for permafrost, *J. Geophys. Res., 84*, 2881, 1979.

Fox, J.L. and A. Dalgarno, Ionization, luminosity, and heating of the upper atmosphere of Mars, *J. Geophys. Res., 84*, 7315, 1979.

Hanson, W.B., et al., The Martian ionosphere as observed by the Viking retarding potential analyzers, *J. Geophys. Res., 82*, 4351, 1977.

Gordy, W. and R.L. Cook, *Microwave molecular spectra*, New York: Wiley, 1984.

Junge, C.E., *Air Chemistry and Radioactivity*, Academic Press, New York, 1963.

Jursa, A.S., ed., *Handbook of geophysics and the space environments*, Hanscom Air Force Base, Bedford, MA, 1985.

Liebe, H. J., Modelling attenuation and phase of radio waves in air at frequencies below 1000 GHz, *Radio Sci., 16*, 1183, 1981.

Liebe, H.J., An updated model for millimeter wave propagation in moist air, *Radio Sci., 20,* 1069, 1985.

Nier, A.O., and M.B. McElroy, Composition and structure of Mars' upper atmosphere: Results from the neutral mass spectrometers on Viking 1 and 2, *J. Geophys. Res., 82,* 4341, 1977.

McElroy, M.B., and J.C. McConnell, Atomic carbon in the atmospheres of Mars and Venus, *J. Geophys. Res., 76,* 6674, 1971a.

McElroy, M.B., and J.C. McConnell, Dissociation of CO_2 in the Martian atmosphere, *J. Atmos. Sci., 28,* 879, 1971b.

McElroy, M.B., T.Y. Kong, and Y.L. Yung, Photochemistry and evolution of Mars' atmosphere: A Viking perspective, *J. Geophys. Res., 82,* 4379, 1977.

National Oceanic and Atmospheric Administration, U.S. Standard Atmosphere - 1976, adopted by the United States Committee on Extension to the Standard Atmosphere, Washington, U.S. Govt. Print. Off., 1976.

Nier, A.O., and M.B. McElroy, Composition and structure of Mars' upper atmosphere: Results from the neutral mass spectrometers on Viking 1 and 2, *J. Geophys. Res., 82,* 4341, 1977.

Owen, T., The composition and early history of the atmosphere of Mars, in *Mars,* eds. H.H. Kieffer, et al., Univ. Arizona Press, 818, 1992.

Smith, E.K., Centimeter and millimeter wave attenuation and brightness temperature due to atmospheric oxygen and water vapor, *Radio Sci., 17,* 1455, 1982.

Ulaby, F.T, R.K. Moore, and A.K. Fung, Microwave Interaction with atmospheric constituents, in *Microwave Remote Sensing: Active and Passive, Vol 1: Microwave Remote Sensing Fundamentals and Radiometry,* ed. by F.T. Ulaby et al., Addison-Wesley Publishing Company, Inc., 256, 1981.

U.S. Dept. of Commerce, Microwave Spectral Tables, Vol. V. *Spectral Line Listing,* National Bureau of Standards Monograph 70, 1968.

Van Vleck, J.H., Theory of absorption by uncondensed gases, *in Propagation of Short Radio Waves,* Vol. 24, *IEE Electromagnetic Wave Series,* Ed. by D. E. Kerr, pp. 646, London: Peter Peregrinus Ltd, 1987.

Waters, J.W., Absorption and emission by atmospheric gases, *in Astrophysics-Part B: Radio Telescopes,* Vol. 12, *Methods of Experimental Physics,* pp. 142, Ed. by M.L. Meeks, Academic Press, 1976.

Yung, Y.L., D.F. Strobel, T.Y. Kong, and M.B. McElroy, Photochemistry of nitrogen in the Martian atmosphere, *Icarus, 30,* 2641, 1977.

5. Martian Dust Storms and Their Effects on Propagation

5.1   Introduction

One of the most remarkable features of Mars is the dust storm. Thus, there are some concerns about dust storm effects on radio wave propagation. We know that Mars has only one third of Earth's surface gravity and that the Martian surface atmospheric pressure is about 1/100 that of Earth. The wind speed at Mars is not significantly larger than at Earth. Thus, a question is how the wind at Mars can frequently raise enough dust to form a storm. To address this question, some wind-tunnel experiments have been performed.

On Earth particles around 0.08 mm in diameter are those most easily moved by the wind [Bagnold, 1941]. Above this diameter, larger wind speeds are required to overcome the inertial resistance due to the particle mass. The threshold velocities on Mars are much larger than those on Earth because of the thinner atmosphere. A wind-tunnel study [Greeley et al., 1980] showed that the optimum size for particle movement on Mars is near 0.1 mm, close to the size for Earth. Figure 5-1 shows that the threshold shear velocities (V_t) required to move the 0.1-mm particles range from 1.4 m/s for a 10-mb atmospheric pressure to 2.1 m/s for a 5-mb atmospheric pressure. On Earth threshold velocities at the optimum size are close to 0.2 m/s.

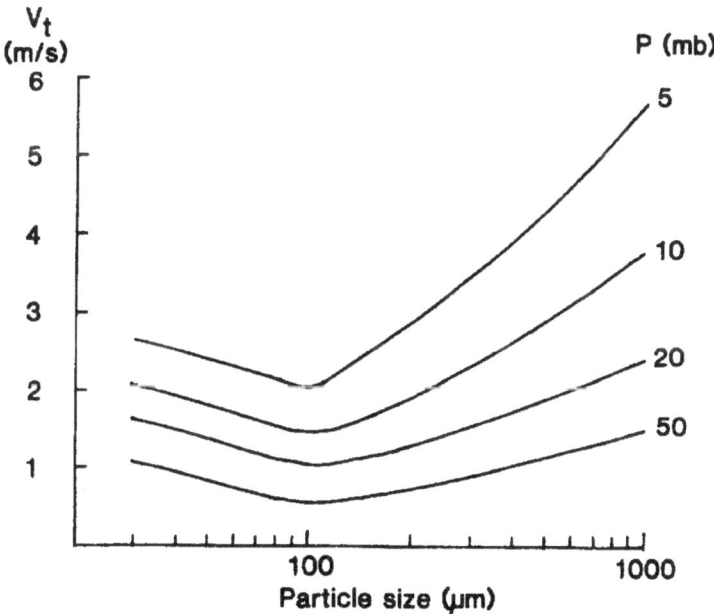

Figure 5-1. Threshold Velocities for Initiation of Particle Movement as a Function of Particle Diameter. The data are from wind-tunnel studies at four pressure levels representative of conditions at the Martian surface (adapted from Greeley et al. 1980).

There is a long history of Mars dust storm observations using terrestrial telescopes. Data on earlier storms have been reviewed by Slipher [1962] and Capen [1971]. Observations after 1971 are particularly well documented in Martin [1976] and Briggs et al. [1979]. Global or great dust storms are most common when Mars is near perihelion. Yellowish dust clouds were extensively photographed during the oppositions of 1920–1922. This period of maximum insolation occurs in late southern spring near L_s ~250°, where L_s is the aerocentric longitude of the Sun as measured in a Mars-centered fixed coordinate system, often used as an angular measure of the Mars year (L_s = 90°, 180°, and 270° corresponding to the beginning of southern winter, spring, and summer, respectively).

Spacecraft such as Mariner 9, Vikings 1 and 2 , Mars Pathfinder, and MGS, have provided direct observations of Martian dust storms. Most of these dust storms occur in spring and summer in the southern hemisphere, in particular along the periphery of the South polar cap. When seasonal heating of the surface and atmosphere induces convection of sufficient strength to raise dust from the surface, the large temperature contrasts drive thermal circulation radially away from the higher ice covered area. These downslope winds are a possible source for dust storms. The great dust storms typically originate as localized dust clouds in the southern temperate zone. These storms expand slowly over a few days, after which they may either decline or rapidly intensify into global dust hazes that may obscure the entire planet. Most of the storms move with velocities of 14–32 m/s and dissipate within a few days. Afterward, the atmosphere usually takes a few weeks or months to become clear. Dust storms can reach as high as 60 km.

Figure 5-2 shows a local dust storm near the edge of the south polar cap imaged by Mariner 9. This fascinating image shows dust swirling over a large area. Martian global dust storms tend to start in the southern hemisphere with a local dust storm, such as the one shown here. Local dust storms seem to be swept into huge storms that envelope the entire planet, as was discovered by the Mariner 9 mission in 1971 and Viking missions to Mars in the 1970's. During 1977 the Viking spacecraft, both orbiters and landers, made extensive observations of dust activity. Between the two global dust storms, numerous local dust storms were observed. Figure 5-3 shows one of the local dust storms observed by the Viking orbiter camera.

Global dust storms do not seem to occur every Martian spring or summer, however. The action of sand carried by winds during Martian global dust storms makes a great contribution toward wearing down rocks on the Martian surface. Martian dust storms are generally divided into three types:

(1) Local dust storms—clouds and hazes with a spatial dimension < 2000 km

(2) Regional dust storms—clouds and hazes with a spatial dimension > 2000 km

(3) Planet-encircling—dust storms that encircle the whole planet at some latitude

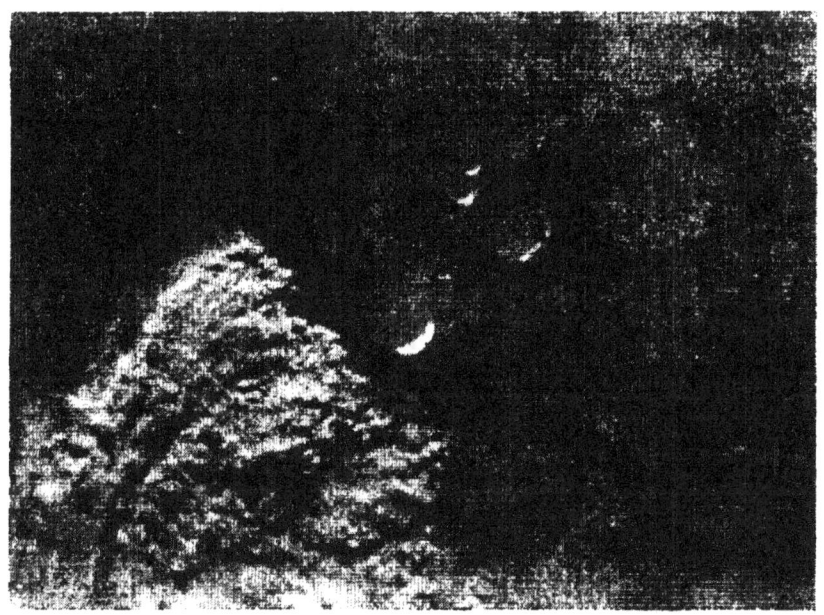

Figure 5-2. A Local Dust Storm Observed by Mariner 9 at the Edge of the South Polar Ice Cap, Just Visible at the Lower Right. The time is that of perihelion, L_s = 250°.

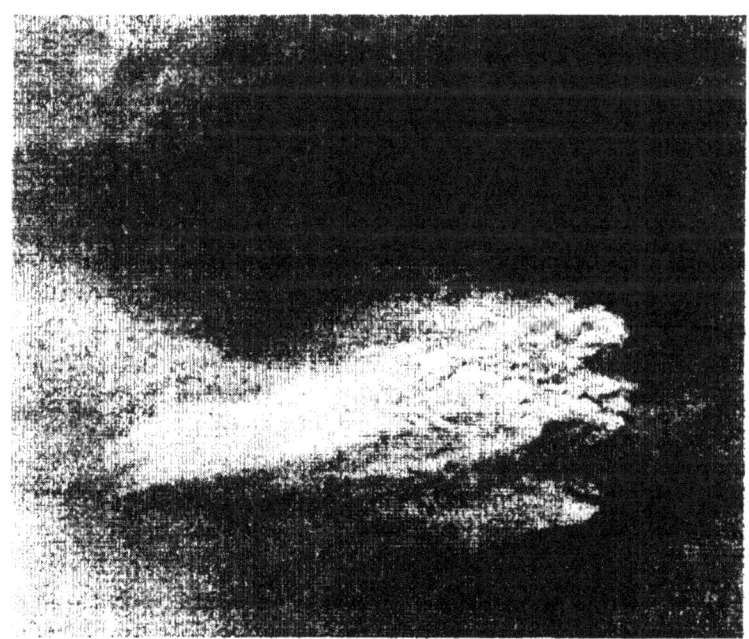

Figure 5-3. A Local Dust Storm in the Solis Planum Region at L_s = 227°. The season is mid-southern spring, between the two 1977 global storms.

5.2   Local and Regional Dust Storms

Local and regional dust storms are relatively common on Mars. They tend to occur in areas of high topographic and/or high thermal gradients (usually near the polar caps) where surface winds would be strongest. It is believed that large temperature differences between the icy polar regions and warmer dark regions of lower latitudes drive the polar dust storms. This type of storm is usually several hundreds of kilometers in extent and is located near the edge of the south polar cap. Some local storms grow larger, others die out. The Martian north polar cap is another region of dust storms.

Observations of the Sun using the camera on Mars Pathfinder showed more dust in the Martian atmosphere than was expected. In fact, the amount of dust was comparable to Viking observations during clear (non-dust-storm) times. The measurements were made by taking images of the Sun in different colors and with the Sun at different elevations in the sky. As the Sun went lower in the sky, the light passed through more dust, becoming fainter and fainter, and indicating the amount of dust. Martian dust includes magnetic, composite particles, with a mean size of one micron.

On April 12, 1998, the Mars Orbiter Camera (MOC) of the Mars Global Surveyor detected a large local dust storm in the original Viking Lander 1 region as shown in Figure 5-4. This view, a map projection, shows an image area of about 310 km wide by 290 km, at a scale of 300 m (985 ft) per pixel. A well-developed local dust storm dominates this view of the planet. Plumes from the storm suggest that the wind is blowing from the lower left towards the upper right. The slightly dark zone around the dust cloud may be a surface that has been swept clean of a fraction of the mobile dust. The dust cloud obscures most of the landing site as seen in this image.

The MOC detected a large, regional dust storm in Noachis Terra on November 26, 1997. The dust storm is the large cloudy feature near the southern polar frost cap that can be seen at the bottom of Figure 5-5. The 1997 dust storm was different from that of 1977. The 1977 storm was first observed at $L_s = 205°$ in the Thaumasia region; it expanded rapidly through the southern hemisphere and then into the northern hemisphere, whereupon dust was observed at both Viking landing sites. During 1997 the southern hemisphere remained clear until after $L_s = 220°$. Subsequently, considerable dust was seen over and around the cap as well as in Noachis. This dust storm never became global, but it was large enough (greater than 2,500 km (1,550 miles) in size) that it would have covered a third of the continental United States. MOC monitored this big dust storm into December 1997. It was most intense between November 26 and December 2, then faded into a haziness that persisted for several weeks. This effect was detected by the Mars Global Surveyor spacecraft during its aerobraking passes [Malin et al., 1998].

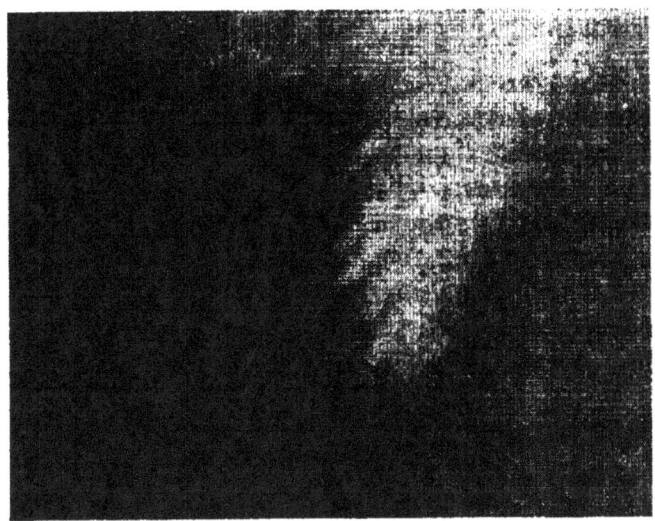

Figure 5-4. Dust Storm Picture Taken by MGS Orbiter Camera (MOC) During Orbit 235
Observations at 22.48°N, 47.97°W. The spacecraft was about 640 km from the surface
with a view angle of 31.64°.

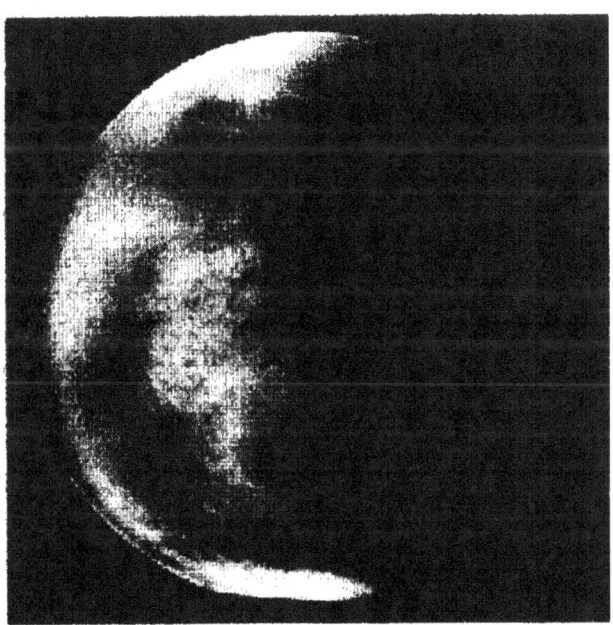

Figure 5-5. A Regional Dust Storm Observed by MGS Orbiter Camera (MOC) from Orbit 50
and L_s = 251° (Malin et al., 1998). The Storm extended from 25° to 60° N latitude and
from 15°W to 40°E longitude, a distance of 2500 km.

5.3   Global Dust Storms

These storms are also called great dust storms or planet-encircling dust storms. We have listed the observed great storms in Table 5-1 [Zurek, 1982; Martin, 1984]. Before the 1971, all observations were made by terrestrial telescopes during each of the most favorable oppositions. Although local dust storms have been observed on Mars during all seasons, the global dust storms have not been observed to originate during northern spring or summer. Global dust storms have a profound impact on Martian environment evolution, such as dust deposits, aeolian processes, sediment features, atmospheric composition, atmospheric circulation, etc. [Christensen, 1986; Greeley et al., 1992; Haberle et al., 1982; Owen et al., 1977; Pollack et al., 1977; Zurek et al., 1992].

Table 5-1. Martian Great Dust Storms

Year	Observation	L_s (°)	Initial Location
1909 (Aug)	Earth		
1911 (Nov)	Earth		
1922	Earth	192	
1924 (Oct)	Earth		
1924 (Dec)	Earth	237	Isidis Planita
1939	Earth		Utopia
1941 (Nov)	Earth		South of Isidis
1943	Earth	310	Isidis
1956	Earth	250	Hellespontus
1958	Earth	310	Isidis
1971 (July)	Earth	213	Hellespontus
1971 (Sept)	Earth, Mariner 9	260	Hellespontus
1973	Earth	300	Solis Planum
1977 (Feb)	Viking	205	Thaumasia
1977 (June)	Viking	275	
1979	Viking	225	

Two global dust storms occurred in 1977 and were extensively observed by the two Viking landers and two orbiters [Kahn et al. 1992; Zurek, 1982]. The first storm developed in the Thaumasia region at $L_s = 205°$. Initially, the storm was seen as a large but localized disturbance resembling other local storms. By the second day, the storm had spread to cover an area thousands of kilometers wide. Visibility improved in 6 or 7 weeks. The second storm apparently developed at $L_s = 275°$ in the same area of the planet as the first storm. It seemed to be a more intense storm than the first one. Figure 5-6 shows the expansion process of the second 1977 storm [Kahn et al., 1992].

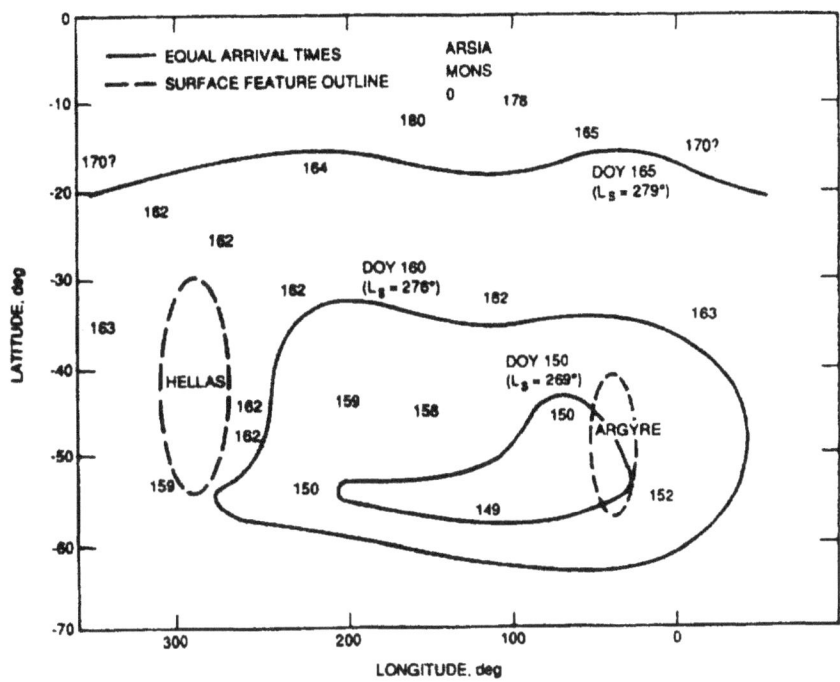

Figure 5-6. Expansion of the June 1977 Storm, as Derived from Viking Orbiter Visual Imaging. Arrival dates of maximum opacity change are shown in day of year, 1977; areocentric longitude Ls is given in parentheses (adopted from Kahn et al., 1992).

A number of local dust storms were observed between the clearing of the first storm and the start of the second. The two Viking landers carried cameras capable of determining the line-of-sight extinction opacity of the atmosphere. Figure 5-7 shows visible optical depth measured by Viking Lander 1 for the two storms. The opacity of a global haze inferred from surface pressure data [Zurek, 1981] has values similar to the optical depths derived from Sun-diode measurements [Colburn et al., 1989]. During the same period the Martian surface pressure had obvious increases that can be correlated with the opacity changes resulting from direct atmospheric heating by the dust.

5.4   Effects on Radio Wave Propagation

5.4.1   Dust Storm Parameters

Mars dust is thought to consist of basalt and montmorillonitic clay. Clear atmosphere corresponds to a background aerosol loading of optical depth ranging from 0.3 to 0.5, at a wavelength of 0.67 mm, while during the most intense portions of the global storms the opacity was found to increase to 4.0–5.0. A local storm generally has a spatial extent of several hundred kilometers. A great dust storm can have a size as big as the state of Texas, or even cover half the planet. Although a storm can reach as high as 50-km altitude, its cloud can be equivalent to a slab with 10-km thickness [Smith and Flock, 1986] because the dust density decreases with increasing altitude. The total amount of dust along a ray path is

$$\int_0^\infty N_0 e^{-\frac{z}{H_d}} dz = N_0 H_d \tag{5-1}$$

where H_d is the dust scale height (~10 km) and N_0 is the surface dust density.

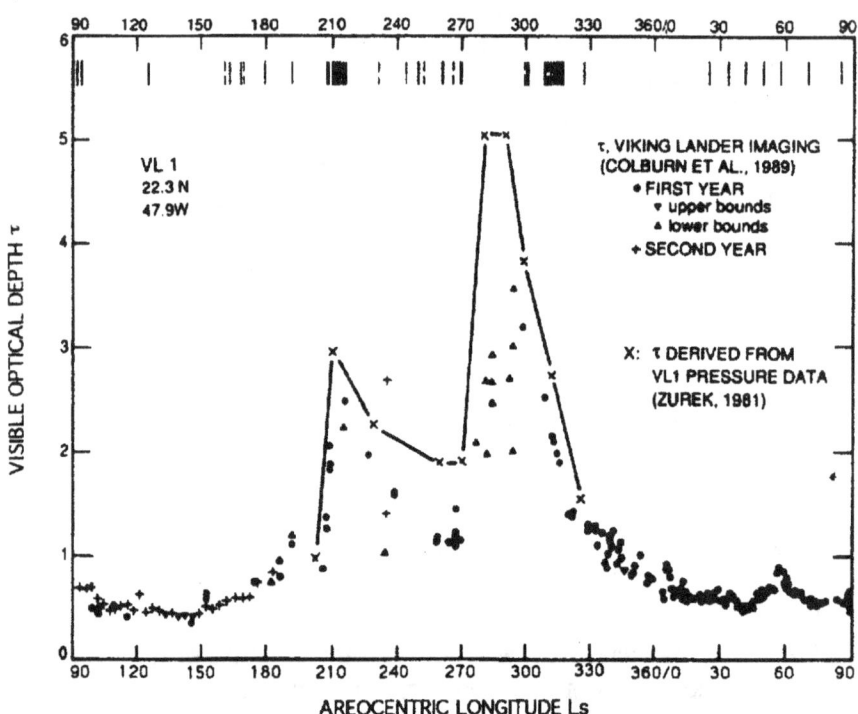

Figure 5-7. Visible Optical Depths Derived from Viking Lander 1 Measurements (Colburn et al., 1989). Line indicates the opacity of a global haze inferred from Viking Lander 1 surface-pressure data (Zurek, 1981). Initial appearance or detection of all dust storms is indicated by vertical bars at the top.

Dust size distribution has been modeled using a modified gamma function [Toon et al., 1977; Hunt, 1979]:

$$N(r) = cr^\alpha \exp[-(\alpha / \gamma)(r / r_m)^\gamma] \tag{5-2}$$

where c is a normalization constant, $\alpha = 2$, $\gamma = 0.5$, and r_m is mean particle radius. Toon et al. [1977] found that $N(r)$ remained reasonably constant throughout the period of the 1971–1972 storms. There is very limited information about dust size. On the basis of the study of Pollack et al. [1979] in assessing the Martian atmospheric opacity during both clear periods and during dust storms, the average dust size is estimated to be on the order of 1–4 μm. Currently we do not understand why Martian dust has such small average size, as compared with Earth. This size of particle is extremely difficult to set into motion by wind, as shown in the experimental results of Greeley et al. [1980].

The refractive index of Martian dust has also been derived based on knowledge of terrestrial dust. Actually the real and imaginary parts of the mean permittivity ($\varepsilon = \varepsilon' + j\varepsilon''$) of dust particles depend on the radio wave frequency, and on the composition, size, and moisture of the dust. In Table 5-2, we list some values that were used in the previous studies.

Table 5-2. Dielectric Permittivity Index of Dust Particles

Index ε	10 GHz[*]	10 GHz Clay	10 GHz Sand	S Band[**]	32 GHz Clay[***]	8.8 GHz Clay[***]	Dust at 20μm[***]	Dust at 2 μm[***]
ε'	4.56 (+0.11, −0.24)	7.42 (+1.73, −1.22)	3.35 (±0.03)	4.56	2.5	2.5	2.0	3.0
ε''	2.51 (+0.074, −0.066)	1.119 (+0.597, −0.437)	0.042 (±0.02)	0.251	0.06	0.02	0.4	0.1

[*]   Ghobrial [1980]

[**]   Goldhirsh [1982]

[***]   Smith and Flock [1986]

5.4.2 Radio Wave Attenuation through Dust Storms

Sand or dust particles can cause attenuation of radio waves through the scattering and absorption by particles. When the particle size is smaller than the wavelength, Rayleigh scattering theory applies. When the particle has a size larger than the wavelength, we should use Mie scattering theory to calculate effective refractive index [Flock, 1987].

For terrestrial sand or dust storms, the visibility, V, is often used to describe the distance at which a mark disappears against the background. Storms usually have a visibility of 10 m or less, with a minimum of 3.8 m, and can reach a height of 1 km or more [Bagnold, 1941; Ghobrial, 1980; Ansari and Evans, 1982]. Dust particles have an average size of 10 to 20 μm, with the largest in a range of 80–300 μm. For an extreme case, with a particle number density, N_T, of $10^8/m^3$, and dust material mass density, ρ, 2.6×10^6 g/m^3, mass loading can reach 40–60 g/m^3. At S-band (10^a cm wavelength), over a 20-km one-way path, attenuation can be as great as 13 dB. X-band (10 GHz) attenuation can reach 44 dB over the same path. Thus, in desert areas, sand storms can cause a problem for domestic-satellite services. Chu [1979] and Goldhirsh [1982] have summarized the studies of the effects of Earth dust storms on radio wave propagation. The microwave attenuation $A(\lambda)$ is given by:

$$A(\lambda) = \frac{189r}{\lambda V}\left[\frac{3\varepsilon''}{(\varepsilon'+2)^2 + \varepsilon''^2}\right] \tag{5-3}$$

where $A(\lambda)$ is in dB/km, V is visibility in kilometers, r is particle radius in meters, and λ is wavelength in meters.

Smith and Flock [1986] performed the first study of radio wave propagation through Martian dust. The computation method is essentially that of Chu [1979] but uses observed maximum optical depth ($\tau = 6$) at 0.67 micron wavelength rather than visibility for Martian dust storms. Under an assumption that a Martian dust cloud is equivalent to a slab with a scale height H_0 (about 10 km) thickness, attenuation may be expressed as

$$A(\lambda) = 54.62 \frac{r\tau}{\lambda} \left[\frac{3\varepsilon''}{(\varepsilon'+2)^2 + \varepsilon''^2} \right] \tag{5-4}$$

where τ is the optical depth, which is a measure of the dust cloud opacity and is defined in the following. For vertical illumination, $I = I_0 e^{-\tau}$, I is the signal intensity at the surface, while I_0 is the intensity with no attenuation. Smith and Flock [1986] found that for a worst case estimate (particle size of 20 μm), the total attenuation for 32 GHz (Ka-band) is about 1.3 dB, relatively small. However, they did not explain why there is a significant difference between Mars' dust attenuation and Earth's, even though both values of N_T and ρ for dust storms are similar in both planets.

When a distribution of particle size is available, we can use another type of expression [Goldhirsh, 1982]:

$$A(\lambda) = \frac{1.029 \times 10^6\, \varepsilon''}{\lambda[(\varepsilon'+2)^2 + \varepsilon''^2]} \sum_i N_i r_i^3 \tag{5-5}$$

where N_i is the number of particles with radii between r_i and $r_i + \Delta r_i$ per m^3. Actually, the term $\sum N_i r_i^3$ gives the dust mass loading. Using a normalized particle-size probability distribution function from Toon [1977], we can integrate this equation to get a total attenuation effect as

$$A(\lambda) = \frac{3.087 \times 10^6\, \varepsilon''\, N_T}{\lambda[(\varepsilon'+2)^2 + \varepsilon''^2]} \int_0^\infty N(r) r^2 dr \tag{5-6}$$

where N_T is the total particle density in #/m^3 and $N(r)$ is the normalized particle number density at radii between r and $r + \Delta r$ as given in equation (5-2). We can also use the mean particle radius $\bar{r}$, to calculate attenuation:

$$A(\lambda) = \frac{1.029 \times 10^6\, \varepsilon''}{\lambda[(\varepsilon'+2)^2 + \varepsilon''^2]} N_T \bar{r}^3 \tag{5-7}$$

where $\bar{r}$ is obtained through an integration over all sizes of particles in $N(r)$.

As a comparison, Table 5-3 lists dust storm parameters and recalculated radio wave attenuation on Earth and Mars. We find that the key difference in parameters is particle size. Because $A(\lambda) \propto N_T r^3$, a 10 times difference in size (10–40 µm at Earth versus 1–4 µm at Mars) results in a three-order-of magnitude difference in attenuation. A 3-dB attenuation at Mars is a worst case estimate for 32-GHz signals over an equivalent 10-km dust cloud consisting of 10-µm radius particles. If a signal propagates obliquely, the attenuation value should be derived by dividing by $cos\theta$, where θ is the angle between the ray and the vertical direction, or dividing by $sin\psi$, where ψ is the elevation angle. At Mars there is much less mass loading as compared with Earth, but its dust number density is one third that of Earth.

Even though there is no direct optical visibility measurement available at Mars, we can use dust distribution to calculate the parameter

$$V = \frac{5.5 \times 10^{-4}}{N_T r^2} \qquad (5\text{-}8)$$

Using $N_T r^2$ values deduced by Pollack et al. [1977, 1979], we find that Mars has a much greater visibility during its dust storms than the Earth.

The attenuation values at Table 5-3 are for Ka-band. However, we can scale the value into other wavelength bands using $A(\lambda) \sim 1/\lambda$. Shorter wavelength (higher frequency) signals undergo greater attenuation.

Table 5-3. A Comparison of Dust Storm Parameters Between Earth and Mars

	N_T m^{-3}	ρ g/m^3	Mean Size (µm)	Maximum Size (µm)	Visibility (m)	Path Length	Attenuation at 32 GHz	Mass Loading
Earth	10^8	2.6×10^6	30–40	80–300	5.1–3.8	10km	65 dB	40–60 g/m^3
Mars	3×10^7	3.0×10^6	1–10	20	184	10km	3 dB	0.4 g/m^3

5.5   Summary and Recommendations

Even though Mars has a thin atmosphere, Martian winds can frequently generate large dust storms. This happens especially during late spring or early summer seasons in the southern hemisphere, when Mars is at its perihelion. The southern hemisphere is suddenly heated, and a large temperature difference relative to the northern hemisphere is generated. This drives strong winds and dust from the southern hemisphere to the northern hemisphere. This process also has a profound impact on the Mars atmospheric environment and surface features.

A global dust storm has a size at least as big as the continental United States, and can cover most of the planet. Frequent local storms have a spatial extent of several hundred kilometers. Dust storms can reach altitudes as high as 50 km with an opacity of 4–6 (probably even 10). However, it is expected that Martian dust storms have relatively small attenuation effects on radio wave propagation because of the small dust particle sizes. At worst, Martian dust storms could have a

3-dB attenuation at Ka-band. For optical communication, attenuation will be much larger. Martian dust has a mean size of 1–10 μm, at least four times smaller than dust on Earth. We do not know why Martian winds can lift such small particles, while wind tunnel experiments show that particles with a size of 100 μm are the most favorable candidates.

To perform an accurate estimate of Martian dust storm attenuation on radio waves, we need detailed information, such as dust particle size distribution, composition, dielectric permittivity, and dust cloud dimensions. So far, we only have very limited knowledge about these parameters. Current values about particle dielectric permittivity is based on similar materials from Earth. For example, if the mean dust particle size increases to 20 μm, the attenuation will increase eight times. Thus, accurate knowledge of the particle size distribution is critical for attenuation calculations. We suggest an extensive measurement and study on Martian dust storms in future Mars missions.

We recommend that mission designers for any spacecraft landing in the southern hemisphere during the late spring or early summer seasons should consider dust storm effects on telecommunication, especially near the southern polar-hood region. Even though local dust storms can occur at any time during the year, and at any location, the most severe dust storms usually take place near the southern polar-hood region. During this period, optical communication is impossible, and Ka-band communications also suffer significant attenuation. We need to have adequate margins (e.g., 5 dB) for signal transmission.

References

Ansari, A.J. and B.G. Evans, Microwave propagation in sand and dust storms, *Proc. IEE, 129*, Part F, 315, 1982.

Bagnold, R.A., *The physics of blown sand and desert dunes*, Methuen London, 1941.

Briggs, G.A., et al., Viking Orbiter imaging observations of dust in the Martian atmosphere, *J. Geophys. Res., 84*, 2795, 1979.

Capen, C.F., Martian yellow clouds-past and future, *Sky & Teles., 41*, 117, 1971.

Christensen, P.R., Regional dust deposits on Mars: Physical properties, age and history, *J. Geophys. Res., 91*, B3, 3533, 1986.

Chu, T.S., Effects of sandstorms on microwave propagation, *Bell Syst. Tech. J., 58*, 549, 1979.

Colburn, D., et al., Diurnal variations in optical depth at Mars, *Icarus, 79*, 159, 1989.

Flock, W.L, *Propagation Effects on Satellite System at Frequencies below 10 GHz, A Handbook for Satellite System Design*, NASA Reference Publication 1108(02), 1987.

Ghobrial, S.F., The effect of sandstorm on microwave propagation, *Proc. Nat. telecomm. Conf.*, Houston, TX, 2, 43.5.1, 1980.

Goldhirsh, J., A parameter review and assessment of attenuation and backscatter properties associated with dust storms over desert regions in the frequency range of 1 to 10 GHz, *IEEE Trans. Ant. Propaga., AP-30*, 1121, 1982.

Greeley, R.R., et al., Threshold wind speeds for sands on Mars: Wind tunnel simulations, *Geophys. Res. Lett.*, *7*, 121, 1980.

Greeley, R. et al., Martian aeolian processes, sediments, and features, in *Mars*, edited by H.H. Kieffer et al., The University of Arizona Press, Tucson & London, 730, 1992.

Haberle, R.M et al., Some effects of global dust storms on the atmospheric circulation of Mars, *Icarus, 50*, 322, 1982.

Hunt, G.E., On the opacity of Martian dust storms derived by Viking IRTM spectral measurements, *J. Geophys. Res., 84*, 8301, 1979.

Kahn, R.A., et al., The Martian dust cycle, 1017, in *Mars*, edited by H.H. Kieffer et al., The University of Arizona Press, Tucson & London, 1992.

Malin, M.C., et al., Early views of the Martian surface from the Mars Orbiter Camera of Mars Global Surveyor, *Science, 279*, 1681, 1998.

Martin, L.J., 1973 dust storm on Mars: Maps from hourly photographs, *Icarus, 29*, 363, 1976.

Martin, L.J., Clearing the Martian air: The troubled history of dust storms, *Icarus, 57*, 317, 1984.

Owen, T., et al., The composition of the atmosphere at the surface of Mars, *J. Geophys. Res., 82*, 4635, 1977.

Pollack, J.B., et al., Properties of aerosols in the Martian atmosphere, as inferred from Viking lander imaging data, *J. Geophys. Res., 82*, 4479, 1977.

Pollack, J.B., et al., Properties and effects of dust particles suspended in the Martian atmosphere, *J. Geophys. Res., 84*, 2929, 1979.

Slipher, E.C., *The Photographic Story of Mars,* Flagstaff: Northland Press, 1962.

Smith, E.K., and W.L. Flock, Propagation through Martian dust at X- and Ka-band, *TDA Progress Report 42-87*, Jet Propulsion Laboratory, Pasadena, CA, 291, July-September 1986.

Toon, O.B., et al., Physical properties of the particles composing the Martian dust storm of 1971–1972, *Icarus, 30*, 663, 1977.

Zurek, R.W., Inference of dust opacities for the 1977 Martian great dust storms from Viking Lander 1 pressure data, *Icarus, 45*, 202, 1981.

Zurek, R.W., et al., Dynamics of the atmosphere of Mars, in *Mars*, edited by H.H. Kieffer et al., The University of Arizona Press, Tucson & London, 835, 1992.

Zurek, R.W., Martian great dust storms: An update, *Icarus, 50*, 288, 1982.

6. Martian Geomorphologic Effects on Propagation

6.1 Introduction

Mars has very complicated geomorphologic (or physiographic) features. These ground structures will affect the radio wave propagation, especially for low-elevation angle propagation. The surface features will affect wave propagation in the following ways [Carr, 1981; CCIR, 1986a and b; Simpson et al., 1992; Goldhirsh and Vogel, 1998]:

1. Multipath at low elevation angle due to reflection and diffraction by terrain and rocks [Vogel and Goldhirsh, 1988]; the ray can arrive at the receiver through more than one path, causing signal fading and attenuation. These features will affect land mobile telecommunication on Mars.

2. Surface reflection, surface roughness, tilt angle, albedo, reflectivity, etc.

3. Atmospheric and climate features associated with surface physiographic structures: fog, water clouds, and dust storms in certain regions.

Reflected radio wave intensities at the Martian surface are strongly dependent on the surface material composition and roughness. There are some changes in the dispersion, polarization, shape, and strength of radio signal echoes due to surface absorption and scattering. In turn, we can use these features to study Martian surface structures. They include surface roughness, reflectivity, small-scale structure size, etc. Surface material properties (reflectivity or dielectric constant) may be inferred from signal changes. For a smooth interface and a normal incident wave, the reflection coefficient, R, is simply related to the dielectric constant ε by

$$R = \left[\frac{1 - \sqrt{\varepsilon}}{1 + \sqrt{\varepsilon}} \right] \qquad (6\text{-}1)$$

where ε can have an imaginary part, which is responsible for loss and absorption. In general, lower material densities have lower reflectivity for a given composition. For example, a solid rock has $\varepsilon \sim 5$ to 9 ($|R| \sim 0.15$ to 0.25), while typical soils are in the range $\varepsilon \sim 2$ to 3 ($|R| \sim 0.03$ to 0.07). The Martian surface has a diverse structure; depending on location, rms surface tilt angles have been found to vary over the range $0.25°$ to $10°$, while reflectivity covers at least 0.03 to 0.25 [Simpson et al., 1992; Bedrossian et al., 2002].

The geomorphology of Mars has an obvious north-south asymmetry. Much of southern hemisphere is heavily cratered. The density of craters larger than 20 km in diameter is particularly high. In contrast, most of the northern hemisphere is covered by plains that have much smaller crater densities, as shown in Figure 6-1 [Wu, 1978; Carr, 1981; Smith et al., 1998]. Among the most impressive features of the planet are the large volcanoes [Mouginis-Mark et al., 1992]. The three largest volcanoes lie in the Tharsis bulge region (centered on the equator at about 115° W). The tallest one, Olympus Mons, is more than 700 km across, with a summit 25 km above the surrounding plains. All Tharsis volcanoes are enormous by terrestrial standards, at 27 km above the Mars reference surface, which is defined as an atmospheric pressure 6.1 mb level [Malin et al., 1998].

Vast dune fields, various albedo patterns that change with time, wind-eroded hills, and drifts of fine-grained material observed at two Viking Lander and Mars Pathfinder sites are all attributed to aeolian processes. Sand dunes are observed in all region of Mars. The largest area of dunes occurs in a broad belt that partly surrounds the north polar ice cap between 75° and 80° N [Carr, 1981]. The only three sites on Mars viewed close up show surfaces sprinkled with rocks in the centimeter to meter size range. In some areas, bedrock can occasionally be seen [Simpson et al., 1992]. At the Mars Pathfinder site shown in Figure 6-2, the images revealed a rocky plain (about 20 percent covered by rocks) that appears to have been deposited and shaped by catastrophic floods. Large rocks are flat-topped and often perched. Between the rocks are mostly fine-grained (< 100 μm) materials. Soils vary from the bright-red dust to darker-gray material. Their composition is mainly sulfur, iron, magnesium, and silicon [Malin et al., 1998].

There is evidence that in the past a denser Martian atmosphere may have allowed water to flow on the planet. Physical features closely resembling shorelines, gorges, riverbeds, and islands suggest that great rivers once marked the planet. Large channels can also be seen everywhere on the Martian surface [Baker et al., 1992]. Some of these channels are several hundred kilometers long and tens of kilometers across. They show numerous indications of early fluvial activity.

A recent landmark discovery is that MGS imagery shows many gullies that may have been be caused by current sources of liquid water at or near the surface of Mars [Malin and Edgett, 2000]. These gullies appear so fresh that they might have been formed yesterday. Figure 6-3 shows an image made by the MGS camera. The gullies observed on cliffs, usually in crater or valley walls, are made up of a deep channel with a collapsed region at its upper end. Nearly all gullies occur between latitudes 30° and 70°. The lack of small craters superimposed on the channels and apron deposits indicates that these features are geologically young. It is possible that these gullies indicate that liquid water is present below the surface of Mars today. This suggests that Mars may have a significant amount of underground water. The atmospheric pressure at the Martian surface is about 6.1 mb (more than 100 times less than it is at sea level on Earth). Liquid water would immediately begin to boil if it were exposed at the Martian surface. At the downstream end of the gully, water may be quickly evaporated or may flow back underground [Malin and Edgett, 2000].

However, the two most significant geomorphological features on Mars are its polar ice caps and its "grand canyon" system. Some layered deposits lie on both polar regions and also extend outward about 10° in latitude [Thomas et al., 1992]. They consist of water or CO_2 ice. On the east side of Tharsis, and just south of the equator, there is a vast interconnected canyon system. In most places, the canyons are more than 3 km deep and 100 km across. This canyon system is called Valles Marineris [Lucchitta et al., 1992]. These features are distinctively different from the rest of the planet. We describe the effects of the geomorphologic features on radio wave propagation in the following paragraphs.

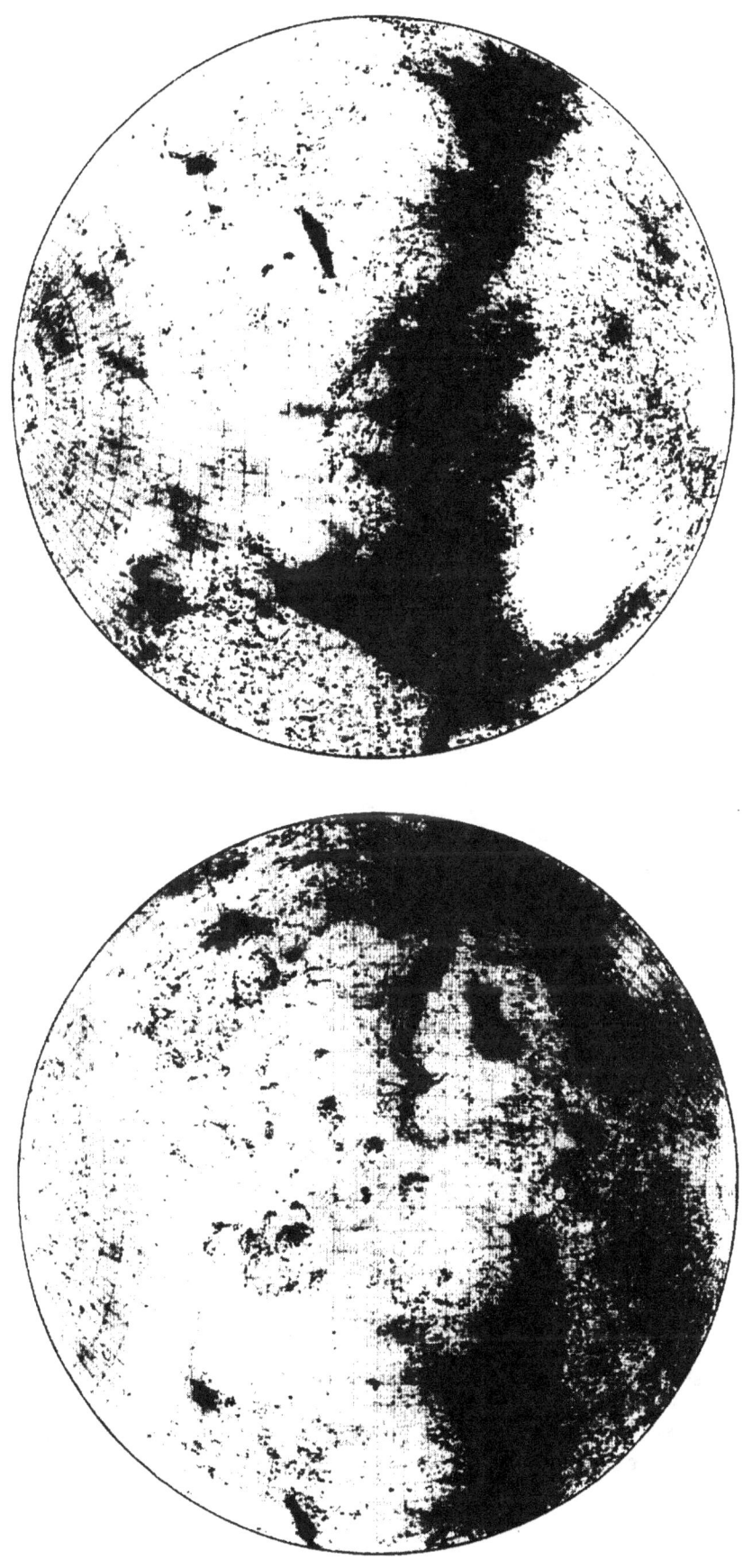

Figure 6-1. Two Lambert Maps Show the Martian Surface Features. The left figure is centered at 120° W; the right at 240° W. There is an obvious north-south asymmetry in the physiography.

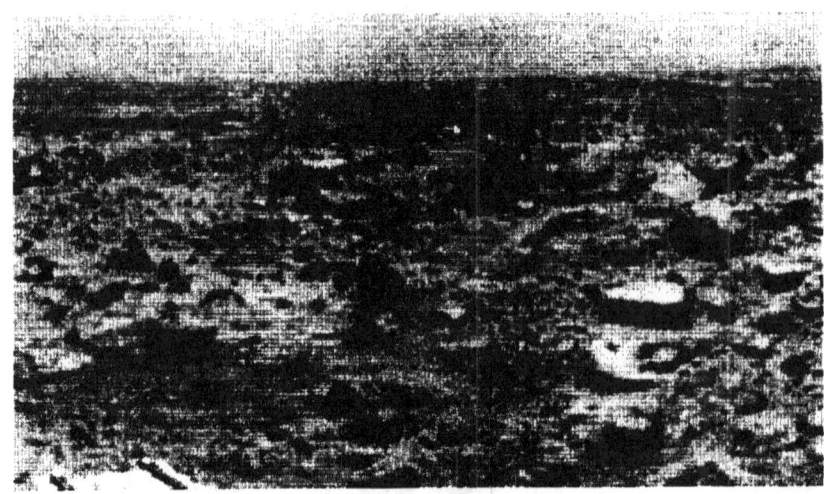

Figure 6-2. Image of Mars Mosaic Showing the Mars Pathfinder Landing Area. The large, prominent rocks comprise the "Rock Garden" which is similar to those of depositional plains in terrestrial catastrophic floods.

Figure 6-3. Image Taken by NASA's Mars Global Surveyor Showing Signs of Water Erosion and Debris Flow. Gullies erode into the wall of a meteor impact crater in Noachis Terra on Mars. Channels and associated aprons of debris are interpreted to have formed by ground water seepage, surface runoff, and debris flow. The scene covers an area approximately 3 km (about 2 miles) wide by 6.7 km (4.1 miles) high, located near 54.8 degrees South by 342.5 degrees West.

6.2 Mars Polar Ice Caps

The Martian polar caps are the first features to be recognized from Earth. Three hundred years ago, Giovanni Cassini (1625–1712) found that both white polar caps changed their sizes with the Martian seasons. We know that Earth's two polar regions (the Arctic and the Antarctic) have huge ice sheets. Ice sheets retreat and advance with the seasons. Similarly, Mars also has ice caps at its polar regions. However, their sizes are much smaller, only 4% of the size of the Earth's Antarctic ice sheet. The Martian polar ice cap also retreats from its maximum (2000 km across) to 300 km (residual cap) in summer, as shown in Figure 6-4 [Thomas et al., 1992].

Both polar ice caps consist of seasonal caps and residual caps. Seasonal caps advance and recede with seasons. During southern winter, Mars is close to aphelion and its orbit moves relatively slowly. Thus, southern winters are colder and longer than northern winters. In contrast, southern summers occur close to the Martian perihelion, so they are relatively hot and short. Only a small remnant cap is left. At its smallest the southern cap is about 350 km across, compared with 1,000 km for the remnant northern cap.

The formation of the caps is not observed because, during the fall, clouds of CO_2 form over the polar hood regions as the CO_2 condenses into the caps. At their maximum extent, the polar caps are roughly circular around the pole. While the southern cap edge can touch the 60° S latitude, the northern one extends to about 65° N. The retreat of the polar caps is nonuniform in longitude but similar from year to year. As the seasonal cap contracts, the characteristic "swirl" texture of the permanent cap emerges. The pattern is caused by preferential removal of frost on equatorward-facing slopes in valleys. The slopes have a roughly spiral pattern in a clockwise direction [Zuber et al., 1998].

The northern residual cap is almost certainly water ice. Brightness temperatures over most of the cap are near 205 K, close to the frost point of water at Martian atmospheric pressure. Solid CO_2 is unlikely to be present in large amounts at this temperature. The southern residual ice cap has a temperature around 160 K, much colder than the northern cap at the same season. This temperature is close to the CO_2 frost point, and thus, the southern cap should predominantly consist of CO_2 ice.

To explain this difference, Pollack et al. [1979] suggested that dust storms started during southern hemispheric summer could play an important role in the residual caps. During that time, the northern cap is formed through CO_2 condensation. Dust in the atmosphere can be seeds for ice condensation, and this would remove a large percentage of the dust grains from the Martian atmosphere. When the southern polar cap is forming (southern winter), there is little dust in the air. Thus, the south polar cap has clean, high-albedo ice that absorbs relatively less solar radiation and is also partly shielded by dust clouds from Sun light. In contrast, the northern cap has "dirty" ice, over which the atmosphere is clear during the southern hemispheric summer. This results in warmer temperatures on the remnant northern summer cap (complete vaporization of CO_2) than on the southern cap (retention of small amounts of CO_2 ice) [Thomas et al., 1992].

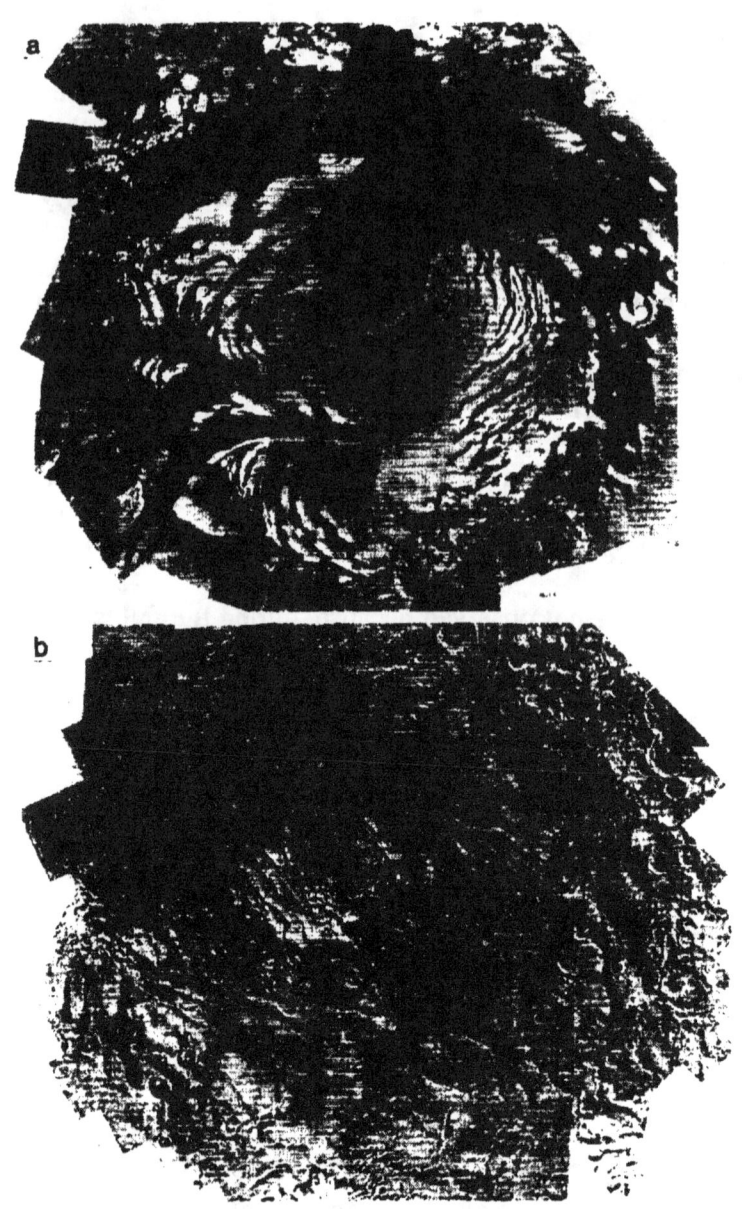

Figure 6-4. Mosaics of Viking Orbiter Images of Polar Regions in the Southern Summer. (a) North polar region; (b) South polar region (from Thomas et al., 1992).

The laser altimeter aboard Mars Global Surveyor produced the first three-dimensional views of the Mars' northern polar cap. This view is shown in Figure 6-5 [Isbell et al., 1998]. During the spring and summer of 1998, as the spacecraft orbited Mars, the altimeter sent laser pulses toward the planet. Very precise elevation data were obtained (1 km spatial resolution, 5–30 m vertical accuracy). Topographic measurements by MGS revealed that the ice cap is about 1200 km across, with a maximum thickness of 3 km. At some sites, the cap is cut by canyons and troughs that plunge to as deep as 1 km beneath the surface. However, some areas are extremely smooth, with only a few meters of changes over many kilometers. It is estimated that the polar cap has a volume of 1.2 million km^3 of water ice. For comparison, this is less than half that of the Earth's Greenland ice cap [Zuber et al., 1998].

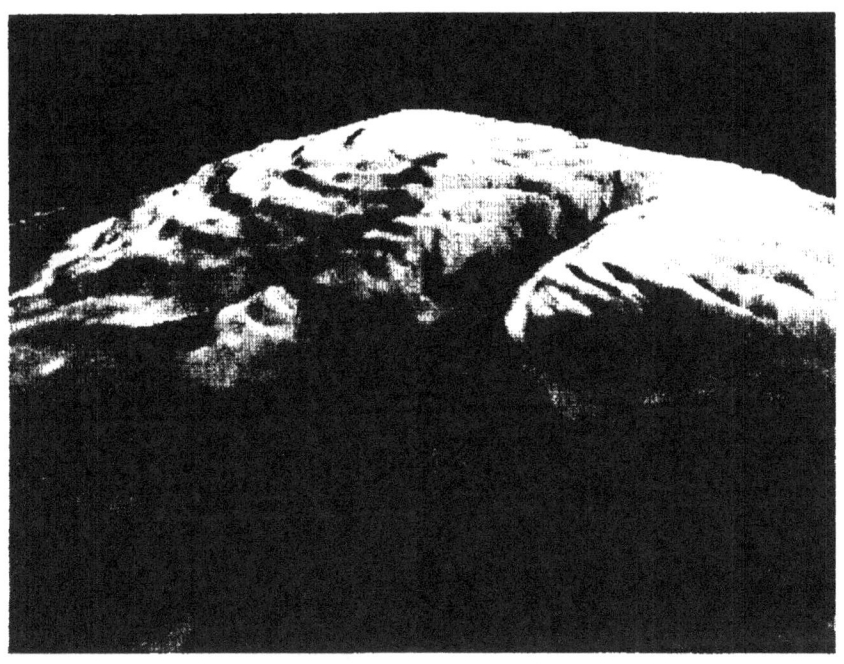

Figure 6-5. Three-Dimensional View of the North Polar Region of Mars from MGS Orbiter Laser Altimeter. The ice cap has a maximum thickness of about 3 km [Isbell et al., 1998; Zuber et al., 1998].

Figure 6-6 shows how diffraction and multipath affect radio wave propagation between a Mars-orbiting satellite and a lander within a polar cap ice trough [CCIR, 1986a and b; Goldhirsh and Vogel, 1998]. As shown in this cartoon, a rover is inside an ice trough, which is about 2 km high and 4 km wide. When the orbiter is at time t_1, even though it cannot directly see the rover, the signals transmitted from the satellite can still reach the rover by a diffracted ray R_1, through the trough edge point T, and a reflected ray R_1' through a reflection from another ice wall. Diffraction is an important possible means of communication between the rover and a satellite beyond the line of sight. On the basis of Huygen's principle, every elementary area of a wavefront can be regarded as a source of secondary spherical waves. Thus, at point T the signal coming from the satellite is a new wave source to propagate to the rover. At times t_2 and t_3, the satellite can directly link to the rover through the rays R_2 and R_3. At the same time, there also are two reflected rays, R_2' and R_3', that arrive at the rover. These reflected rays are delayed in time and phase relative to the direct rays. Assuming that the height of point T on the right ice wall is h_1 (about 2.0 km), the elevation angle of the right side edge relative to the rover is α, while the satellite's elevation angle is θ at time t_1 beyond the line of sight. Also, assuming that the directly received signals R_2 or R_3 (at t_2 or t_3 respectively) at the rover have an intensity E_0,

$$E_{R_2} = E_{R_3} = E_0 \qquad (6\text{-}2)$$

79

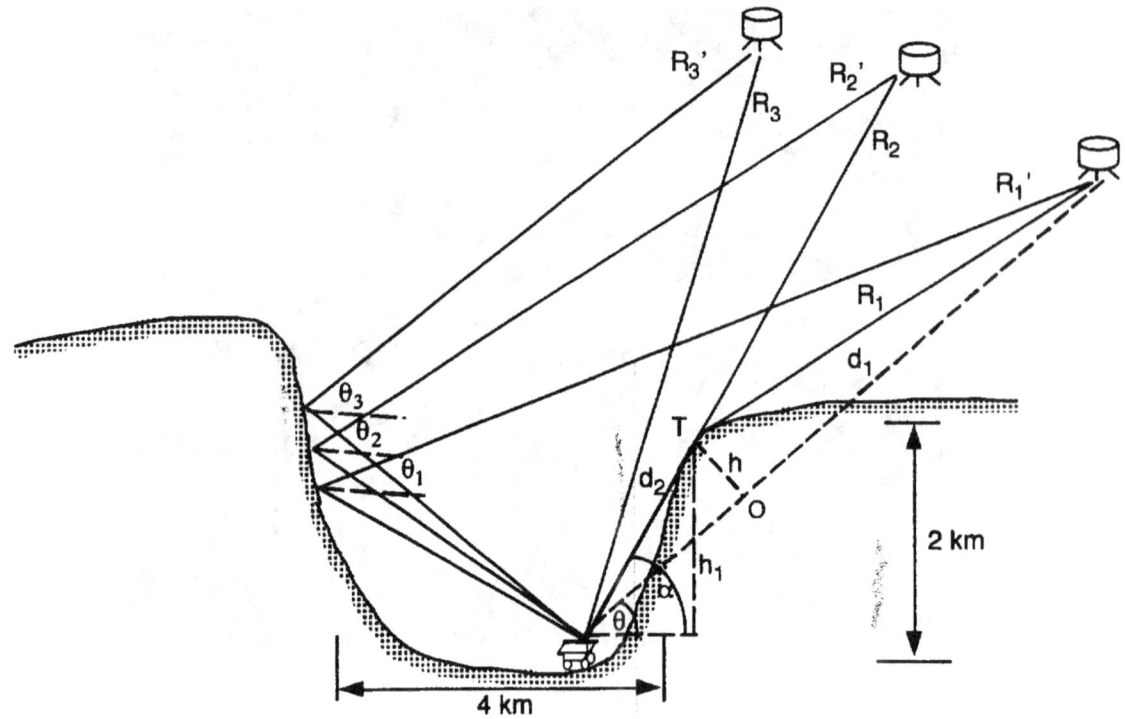

Figure 6-6. Graphic Diagram Showing Radio Ray Paths Between a Rover Within a Canyon and a Satellite Orbiting Mars. Below the line of sight, through ray diffraction and reflection, the rover may still communicate with the satellite, but radio signals will have a higher attenuation than the direct link.

The diffracted signal has an intensity (instead of strength) of

$$10\log_{10}\frac{E_{R_1}}{E_0} = -J(\nu) \tag{6-3}$$

$$E_{R_1} = E_0 10^{-J(\nu)/10} \tag{6-4}$$

where

$$J(\nu) = 6.9 + 20\log(\sqrt{(\nu-0.1)^2 + 1} + \nu - 0.1) \tag{6-5}$$

is diffraction loss (in dB), and the parameter ν is

$$\nu = h\sqrt{\frac{2}{\lambda}(\frac{d}{d_1 d_2})} \tag{6-6}$$

This comes from the attenuation over a single knife-edge obstacle [Deygout, 1966]. In Equation (6-6), $h = \dfrac{sin(\alpha - \theta)}{sin\alpha} h_1$, λ is the signal wavelength, d_1 and d_2 are distances of two ends of the path from the top of the obstacle, and d is total path length. Because $J(v)$ is usually greater than 6.9, the diffracted signal intensity over a canyon edge, E_{R_1}, is less than $0.2 E_0$.

The reflected signals along the paths R_1', R_2', and R_3' have intensities

$$E_{1,2,3}' = \rho E_0 e^{i\phi_2} = (|\rho| e^{i\phi_1})(|E_0| e^{i\phi_2}) = |\rho||E_0| e^{i\Phi} \tag{6-7}$$

where ρ is reflection coefficient, which may be a complex and will not have a magnitude of unity. ϕ_1 is the phase of reflection coefficient, and ϕ_2 is the phase shift corresponding to the path length difference (Δl) between the reflected ray and the direct ray. Thus, the phase angle Φ of the reflected ray relative to the diffracted ray is

$$\Phi = \phi_1 + \phi_2 = \phi_1 + \frac{2\pi\Delta l}{\lambda} \tag{6-8}$$

The signal intensities received by the rover should be a combination of the direct ray (or diffracted ray) E_R and the reflected ray E' [Beckmann and Spizzichino, 1963]. At t_2 and t_3

$$E_{2,3} = E_0 + E_{2,3}' = |E_0| \sqrt{1 + |\rho|^2 - 2|\rho| \cos\Phi} \tag{6-9}$$

At t_1, total field is

$$E_1 = E_{R_1} + E_1' = |E_0| \sqrt{10^{-2J(v)/10} + |\rho|^2 - 2|\rho| 10^{-J(v)/10} \cos\Phi} \tag{6-10}$$

Except for direct measurements, it is almost impossible to calculate a precise value for Φ, even though we know all the geometric parameters, because the path-length difference depends largely on atmospheric effects on the ray paths (curved to some degrees and variable with time).

When $\rho = 1$, we have

$$E_{2,3} = 2|E_0| \cdot sin\frac{\Phi}{2} \tag{6-11}$$

We have maximum and minimum intensities for $E_{2,3}$ of

$$E_{max} = |E_0|(1 + |\rho|) \quad \text{when } \Phi = \pi \tag{6-12}$$

$$E_{min} = |E_0|(1 - |\rho|) \quad \text{when } \Phi = 0 \tag{6-13}$$

For E_1, we have

$$E_{\max} = (|\rho| + 10^{-J(v)/10}) |E_0| \quad \text{when } \Phi = \pi \tag{6-14}$$

$$E_{\min} = (|\rho| - 10^{-J(v)/10}) |E_0| \quad \text{when } \Phi = 0 \tag{6-15}$$

Its intensity depends on both of reflection coefficient and diffraction attenuation.

6.3   Mars "Grand Canyon" Valles Marineris

Just south of the equator, between the longitudes 30° W and 110° W, are several enormous, interconnected canyons, which have been collectively called Valles Marineris. The Valles Marineris is the most spectacular geologic feature on Mars. The canyon is 4000 km long, 150 km wide, and 10 km deep as shown in Figure 6-7. By comparison, the Grand Canyon is 450 km long, 30 km across, at its widest point, and 2 km deep. Most of the interconnected canyons lack indications of fluvial action on their floors. They appear to result from faulting along east-west faults. The canyons are widest and deepest in the central section, between 65° W and 77° W, where three huge parallel troughs are each close to 200 km across [Lucchitta et al., 1992].

Canyon walls usually have steep gradients and stand at great heights relative to their top plateaus. Many canyon walls have slope angles greater than 35°, their limit of mechanical stability. Some walls have two parts, upper parts with steep slopes and lower parts with sharp-crested ridges, which are oriented perpendicular to the canyons. The canyon floors vary considerably from place to place. Where the canyons are narrow, the floors tend to be segmented. Where the troughs are wide, the floors are flatter and better integrated and may include landslide deposits.

Figure 6-8 shows laser altimeter reflectivity measurements from Mars taken by MGS during fall 1997 (L_s = 198-212). The small reflectivity values over the Valles Marineris canyon can be interpreted as a combination of surface geometric albedo and two-way atmospheric transmission [Ivanov and Muhleman, 1998]. The reflectivity (R) is the ratio of returned laser energy to the emitted laser energy. R is affected by the surface albedo (A) of the underlying terrain and extinction of the photons from the laser beam by atmospheric aerosols. R can be expressed as

$$R = A \ e^{-2\tau} \tag{6-16}$$

where τ is the atmospheric opacity of the atmosphere. Using relative surface albedo data from Viking [Colburn et al., 1989], large atmospheric opacity values are obtained, which are interpreted in terms of atmospheric aerosol loading in the canyons. There are large opacity changes with depth within a canyon.

NASA is planning a series of Mars Micromissions for 2003, 2005, and later. One of the possible missions, called Mars Airborne Geophysical Explorer (MAGE), will use an airplane to determine how the canyons of Valles Marineris formed and have subsequently evolved. The planned Mars airplane, known as "Kitty Hawk," will carry a payload of several instruments on a 3-hour,

82

1800-km flight over the canyon. Fully deployed, the 135-kg airplane has a long (9.75 m) wingspan and a comparatively small body (because of lower Martian pressure), as shown in Figure 6-9. Using the airplane, a wide range of gravitational and topographic features can be closely surveyed. The low-drag design will allow Kitty Hawk to fly at a constant pressure altitude, which will vary from 1 km to 9 km from the canyon rim to its floor. To drive the rear-mounted propeller and provide electric power, the aircraft will use a hydrazine-fueled engine. Possibly in 2010, cruise and relay vehicle (CARV) will carry a Mars atmosphere entry vehicle housing the Kitty Hawk airplane from Earth to Mars. After releasing the entry vehicle at Mars, CARV will act as a data relay to collect data from the airplane and relay it to Earth.

Figure 6-10 shows an example of how diffraction over a canyon edge affects communication between the satellite and a Martian airplane [CCIR, 1986a]. Because of the roughness of the rock surface, no multipath (like those in Figure 6-5) caused by reflection from walls is considered in this calculation. A Martian airplane is inside the canyon, flying along the 6.1-mb pressure level. The altitude difference between the airplane and the canyon edge is h_1 (about 1.0 km). The elevation angle of the canyon edge relative to the airplane is α, while the satellite's elevation angle is θ at time t_1 behind the sight. The radio wave R_1 will be diffracted by the canyon edge at point T. Assuming that directly received signals R_2 or R_3 (respectively at t_2 or t_3) by the airplane have an intensity E_0, the diffracted signal will have an intensity

$$E_{R_1} = E_0 10^{-J(v)/10} \tag{6-17}$$

where $J(v)$, v, has the same definitions as shown in Equation (6-4). Parameters h_1, d_1, d_2, d, θ, and α are also defined as in Figure 6-10.

6.4   Summary and Recommendations

Mars has very complicated surface geomorphologic structures. These structures will affect radio wave propagation from surface equipment and from an aircraft to a satellite when the satellite is at low elevation angles. When terrain or rocks block the direct radio wave rays, a lander can sometimes still communicate with a satellite through diffracted and/or reflected rays. The intensities of diffracted wave signals are reduced by at least a factor of 5 relative to direct signals. Reflected signal intensity depends largely on the reflection coefficient of the reflecting materials. Multipath can also cause signal amplitude fading and attenuation due to phase shifts. Sometimes, when reflected signals have a nearly 180° phase difference from direct (or diffracted) signals, the combined signal intensity can be severely attenuated (by ~30 dB). This will be a potential problem for future Mars communication with mobile land sources and with colonies. Multipath fading effects can be modeled by a Rayleigh or Ricean model. Fading mechanisms and distributions of Martian rocks and terrain in various spatial scales need to be studied in the future. Currently we have very limited information about the surface and rock properties. We recommend inclusion of such measurements and experiments in future Mars missions.

Figure 6-7. A Color Image of Valles Marineris, the Great Canyon of Mars; North Toward Top. The scene shows the entire canyon system, over 3,000 km long and averaging 8 km deep, extending from Noctis Labyrinthus, the arcuate system of grabens to the west, to the chaotic terrain to the east. This image is a composite of Viking medium-resolution images in black and white and low-resolution images in color; Mercator projection. The image extends from latitude 0 degrees to 20 degrees S. and from longitude 45 degrees to 102.5 degrees.

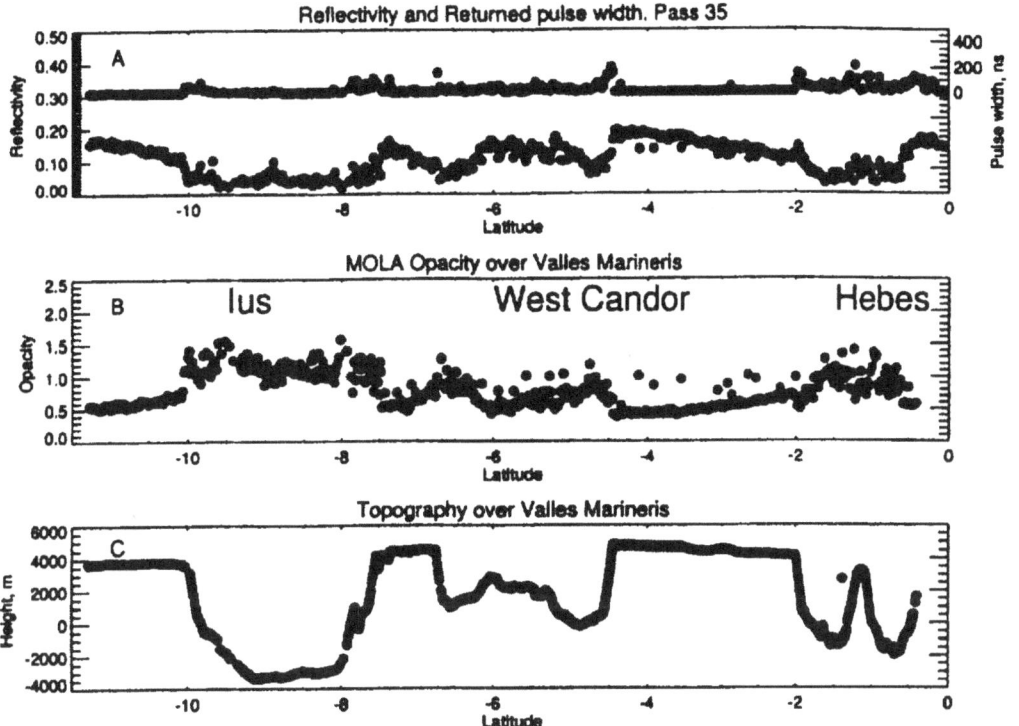

Figure 6-8. Opacity of the Martian Atmosphere Over Valles Marineris Canyon from the Mars Orbiter Laser Altimeter (MOLA) Observations. a) Reflectivity and returned signal pulse width; b) Atmospheric opacity; c) Canyon topography in the latitude range 11° S to 0° N near longitude 77° W [Ivanov and Muhleman, 1998].

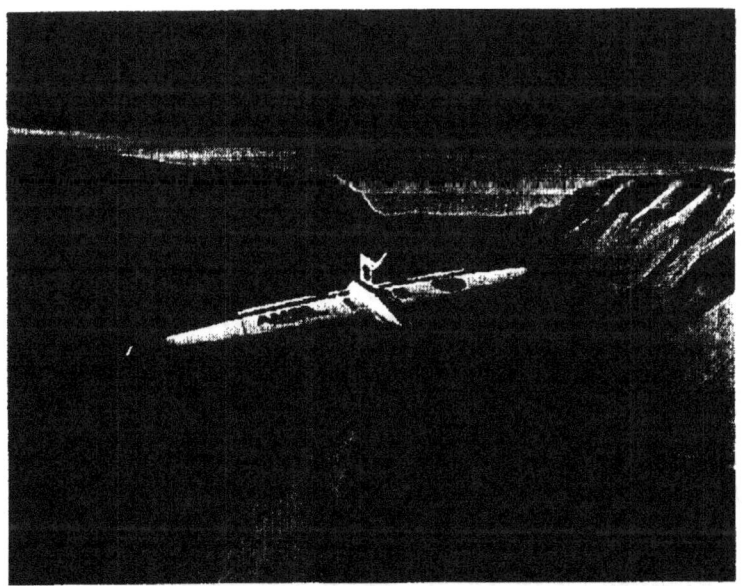

Figure 6-9. Artist's Conception of a Mars Airplane Flying over Valles Marineris.

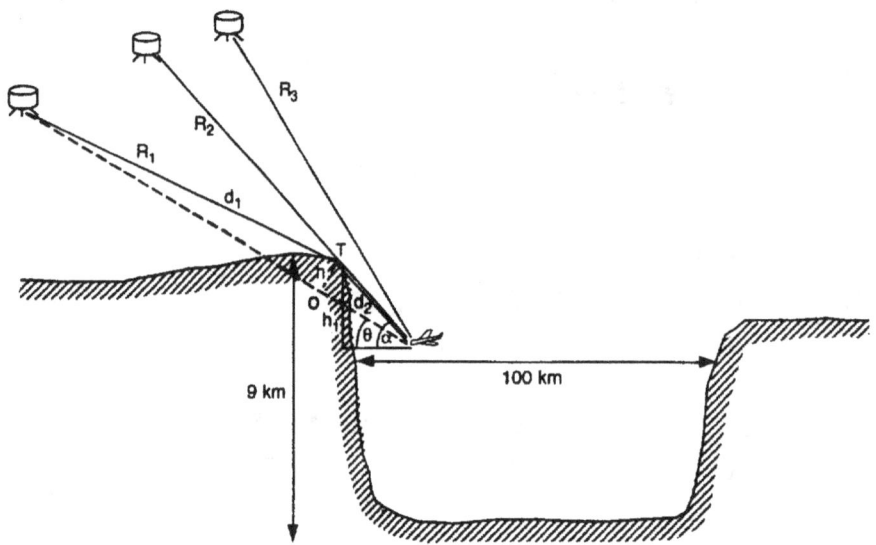

Figure 6-10. Geometric Plot Showing the Ray Paths Between a Satellite and a Mars Airplane. The Mars airplane will fly along a constant atmospheric pressure level. This surface may sometimes be below the canyon edge. However, through refraction and diffraction, the airplane may still maintain communication with satellite.

References

Baker, V.R., et al., Channels and valley networks, in *Mars*, eds. by H.H. Kieffer et al., Univ. of Arizona Press, 493, 1992.

Beckmann, P. and A. Spizzichino, The Scattering of Electromagnetic Waves from Rough Surfaces, New York: Macmillan Co., 1963.

Bedrossian, A., et al., Survey of reflection and scattering measurements of radiowave by Mars surface for communications on and around Mars, *InterPlanetary Network Progress Report*, Jet Propulsion Laboratory, Pasadena, California (in press).

Carr, M.H., *The surface of Mars*, Yale University Press, 1981.

CCIR, Propagation by diffraction, Report 715-2 in Vol. V, *Propagation in Non-ionized Media, Recommendations and Reports of the CCIR*, 1986, Geneva: Int. Telecomm. Union, 1986a.

CCIR, Reflection from the surface of the Earth, Report 1008 in Vol. V, *Propagation in Non-ionized Media, Recommendations and Reports of the CCIR*, 1986, Geneva: Int. Telecomm. Union, 1986b.

Colburn, D., et al., Diurnal variations in optical depth at Mars, *Icarus, 79*, 159, 1989.

Deygout, J., Multiple knife-edge diffraction of microwaves, *IEEE Trans. Veh., Technol. Vol. AP-14*, 480, 1966.

Goldhirsh, J. and W.J. Vogel, *Handbook of Propagation Effects for Vehicular and Personal Mobile Satellite Systems: Overview of Experimental and Modeling Results*, APL-A2A-98-U-0-021, 1998.

Isbell, D., C.M. O'Carroll, and D. Ainsworth, Laser provides first 3-D view of Mars' north pole, Presented at 1998 fall meeting of AGU, San Fracisco, Dec. 6, 1998.

Ivanov, A., and D.O. Muhleman, Opacity of the Martian atmosphere from the Mars Orbiter Laser Altimeter (MOLA) observations, *Geophys. Res. Letts. 25*, 4417, 1998.

Lucchitta, B.K., et al., The canyon system on Mars, in *Mars*, eds. by H.H. Kieffer et al., Univ. of Arizona Press, 453, 1992.

Malin, M.C., and K.S. Edgett, Evidence for recent groundwater seepage and surface runoff on Mars, *Science, 288*, 2330, 2000.

Malin, M.C., et al., Early views of the Martian surface from the Mars Orbiter Camera of Mars Global Surveyor, *Science, 279*, 1681, 1998.

Mouginis-Mark, P.J., et al., The physical volcanology of Mars, in *Mars*, eds. by H.H. Kieffer et al., Univ. of Arizona Press, 424, 1992.

Pollack, J.B., et al., Properties and effects of dust particles suspended in the Martian atmosphere, *J. Geophy. Res., 84*, 2929, 1979.

Simpson, R.A., et al., Radar determination of Mars surface properties, in *Mars*, eds. by H.H. Kieffer et al., Univ. of Arizona Press, 652, 1992.

Smith, D.E., et al., Topography of the northern hemisphere of Mars from the Mars Orbiter Laser Altimeter, *Science, 279*, 1686, 1998.

Thomas, P., et al., Polar deposits of Mars, in *Mars*, eds. by H.H. Kieffer et al., Univ. of Arizona Press, 767, 1992.

Vogel, W.J., and J. Goldhirsh, Fade measurements at L-band and UHF in mountainous terrain for land mobile satellite systems, *IEEE Trans. Anten. Prop., Vol., AP-36*, Jan. 1988.

Wu, S.S.C., Mars synthetic topographic mapping, *Icarus, 33*, 417, 1978.

Zuber, M.T., et al., Observations of the north polar region of Mars from the Mars Orbiter Laser Altimeter, *Science, 282*, 2053, 1998.

7. Propagation Issues for Communication Between Earth and Mars

7.1 Free Space Loss Between Mars and Earth

When the Earth passes between Mars and the Sun (opposition), the minimum distance between the two planets is about 55×10^6 km. In contrast, when the two planets are on opposite sides of the Sun (superior conjunction), their maximum distance is close to 400×10^6 km.

Signal spread loss in free space is defined as L_{FS}

$$(L_{FS})_{dB} = 10 log(\frac{4\pi d}{\lambda})^2 = 20 log(\frac{4\pi d}{\lambda}) \tag{7-1}$$

Using this equation and the distances given above, we have calculated and listed the free space losses for various frequency bands in Table 7-1. Higher frequencies have larger losses, but mission designers can partially compensate for the losses by using higher antenna gain.

Table 7-1. Free Space Losses for Various Frequencies Between Mars and Earth

	Distance d	VHF Band (300MHz)	S Band (3 GHz)	X Band (10 GHz)	Ka-band (32 GHz)
Minimum Distance	55×10^6 km	~237 dB	~257 dB	~267 dB	~277 dB
Maximum Distance	400×10^6 km	~254 dB	~274 dB	~284 dB	~294 dB

The telecommunication system sets up the data transmission rate based on these losses. In order to maintain per-bit energies above the threshold, a lower data rate is usually used for a greater distance (that is, larger loss), while a higher data transmission rate is used for a lesser distance. This system is also set for a small margin (about 5 to 8 dB) for Earth weather degradation. Thus, when very bad weather occurs (heavy rain and clouds), the margin will be exceeded. Because both signal attenuation and background noise temperature increase, reception of high-transmission rate signals becomes difficult. This forces the system to change into a lower data transmission rate.

7.2 Combined Propagation Losses Under Normal and Worst Conditions

Because both Earth and Mars have ionospheres and atmospheres, radio waves suffer some losses in additional to the free space loss when they propagate through these media. At Earth, for 99% of the time, weather conditions are such that the total tropospheric attenuation for Ka-band is about 5 dB for vertical propagation. This loss includes gaseous absorption, rain and cloud scattering, etc. Among these losses, the dominant loss is due to rain scattering and absorption, about 3–4 dB under normal conditions. For the worst case, which occurs about 0.001% of the time, rain attenuation can be as large as 40–50 dB. Another large attenuation source is terrestrial dust storms. For extreme cases, a dust storm can cause a 50–60 dB attenuation at Ka-band.

Fortunately, these dust storms only occur in limited areas in the world, such as the Gobi Desert. Under normal conditions, these storms should only cause an attenuation of less than 3 dB. Also, dust storms are usually separate from rain storms. They almost never occur simultaneously at the same location. Thus, in the worst case (0.001% of the time), a Ka-band signal will suffer a 50 dB loss. A telecommunication system definitely cannot work at a normal transmission rate under this worst condition.

At Mars, the dominant attenuation factor is dust storms. For a worst case (large mass loading), attenuation can be 3 dB or higher at Ka-band. However, this type of storm rarely occurs. Dust storms mostly occur in the southern hemisphere during the spring and summer seasons. Under normal conditions, a storm can cause at most about a 1 dB loss.

At Mars no rain observation has been reported yet. Even though it is possible to have rain, the rain would be so light that it would not cause any significant attenuation to radio waves. It is estimated that total tropospheric losses, including gaseous attenuation, cloud, fog, and tropospheric scattering (scintillation and turbulence), etc., are about 0.4 dB at Ka-band. Thus, under normal conditions, the attenuation combined from a dust storm and the troposphere is about 1.4 to 2 dB for a vertically propagating wave (compared with about 5 dB at Earth). The total attenuation will be about 3.4 dB for the worst case. The attenuation parameters for various frequency bands are listed in Table 7-2.

The Martian ionosphere will have some absorption and scintillation effects on VHF wave transmission, just as the Earth's ionosphere does. Here we have used 0.5 dB for VHF band signal and smaller losses for higher frequency bands. The exact losses are as yet unknown because we do not know the collision frequency and irregularities in the Martian ionosphere. At Earth, these losses are about 3.0 to 10 dB for at 127 MHz. At Mars, this type of loss will be much smaller, because the Martian ionosphere is one order of magnitude thinner than Earth's. VHF band (400–500 MHz) waves have been used for communication between a rover and a lander (Mars Pathfinder) and between a lander and an orbiter. The ionospheric effects on these waves should be further studied.

Another important attenuation factor for Martian surface communication is multipath due to reflections from rocks and canyon walls. Mars Pathfinder and Viking landing areas showed a lot of rocks and hilly structures. The communication between a rover and a base station will be affected by rock distributions and the surface refraction coefficient. Because there have been no experiments yet to measure these parameters on Mars, we must extrapolate from Earth-based experiments. We do not expect that there are any significant differences in attenuation between rocks at Mars and Earth. Goldhirsh and Vogel [1998] have studied multipath effects for canyon and hilly environments. For 870-MHz waves, attenuation has a range of 2–7 dB, while for L-band (1.7 GHz), the attenuation is 2–8 dB. At higher frequencies, higher losses should be expected. Thus, surface rock attenuation is a potentially a large attenuation source.

Table 7-2. Radio Wave Attenuation Around Mars for Various Frequency Bands

	VHF (100–500 MHz)	S-Band (2–4 GHz)	X-Band (10–12 GHz)	Ka-Band (30–38 GHz)
Ionosphere (absorption & scintillation)	0.5 dB	0.15 dB	0.1 dB	0.05 dB
Troposphere (scattering)	0	0	0	negligible
Gaseous	0	0 dB	0 dB	0 dB
Cloud	0	0	0.05 dB	0.1 dB
Rain	0	0	0	0
Fog	0	0	0	0.1 dB
Aerosol (haze)	0	0	0	0.1 dB
Dust*	0.1 dB	0.3 dB	1.0 dB	3.0 dB
Total Vertical Losses	0.5 dB	0.45 dB	1.15 dB	3.35 dB

* Worst case

Figure 7-1 schematically shows all possible Martian communication links from a point of view of wave propagation. Attenuation values for each link at four frequency bands are listed in Table 7-3. For surface-to-surface propagation, we do not know what the actual loss is because there is not yet any rock attenuation experimental data. The total propagation loss between Mars and Earth is free-space loss, plus about an 8-dB atmospheric loss from both planets.

When we calculated total losses, we ignored medium loss through interplanetary space. Actually, interplanetary space is not empty. Propagation of radio waves on deep space paths is affected by solar wind particles, or extended solar corona in interplanetary space. Dust particles in space are believed to be responsible for zodiacal light [Halliday and McIntosh, 1980]. Quoted values of the density of interplanetary dust that have been noted are about 10^{-17} or 10^{-18} g/m^3 (10^{-23} or 10^{-24} g/cm^3). The total attenuation (A) is proportional to mass loading:

$$A = k_a \int \rho dl = k_a \rho l \qquad (7-2)$$

where k_a is the coefficient (= 3×10^{-4} dB/m^2) [Flock, 1981], ρ is constant mass density (= 10^{-17} g/m^3), and l is the path length (= 4×10^{11} m). Because the dust abundance is so low and the particle sizes are so small, the attenuation effect on microwaves due to the dust of interplanetary space (1.2×10^{-9} dB) is negligible.

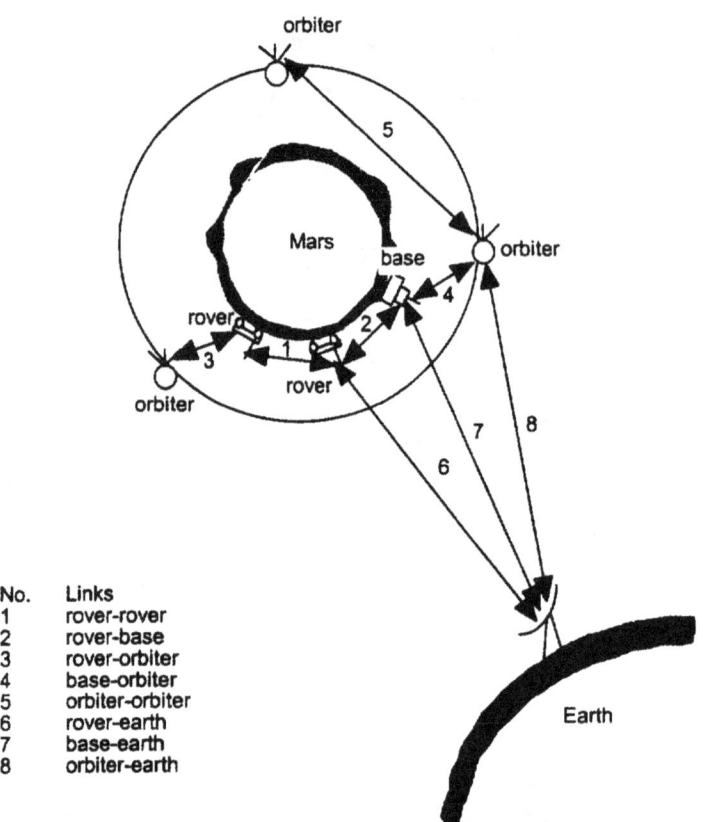

Figure 7-1. Telecommunication Links Around Mars from the Point of View of Radio Wave Propagation. There is a total of eight possible links between Mars and Earth.

No. | Links
1 | rover-rover
2 | rover-base
3 | rover-orbiter
4 | base-orbiter
5 | orbiter-orbiter
6 | rover-earth
7 | base-earth
8 | orbiter-earth

Table 7-3. Attenuation for All Possible Links Between Mars and Earth

Link Type	VHF (100–500 MHz)	S-Band (2–4 GHz)	X-Band (10–12 GHz)	Ka-Band (30–38 GHz)
Rover-Rover*	—	—	—	—
Rover-Base*	—	—	—	—
Rover-Orbiter**	0.1 dB	0.3 dB	1.0 dB	3.4 dB
Base-Orbiter**	0.1 dB	0.3 dB	1.0 dB	3.4 dB
Orbiter-Orbiter**	0	0	0	0
Rover-Earth***	243 dB	263 dB	273 dB	283 dB
Base-Earth***	243 dB	263 dB	273 dB	283 dB
Orbiter-Earth***	240 dB	260 dB	270 dB	280 dB

* Depends on distance and terrain.

** Free space loss excluded.

*** Attenuation at the maximum distance between Mars and Earth, add 17 dB.

The solar corona and the solar wind consist of ionized gases: plasma. The electron density distribution of the solar corona has a strong dependence on the radial distance from the Sun

[Smith and Edelson, 1980]. At 1.1 R_s (solar radius) (1.1 × solar radius [6.96×10^5 km] = 7.66 × 10^5 km), the electron density is 1.25×10^{14} m^{-3}. At 4 R_s (2.78×10^6 km), the density is 1.2×10^{11} m^{-3}, which is less than the peak densities of the Martian ionosphere (2.5×10^{11} m^{-3}) and the Earth's ionosphere ($2.0 \times 10^{12} m^{-3}$). At 1 AU (215 R_s, 1.46×10^8 km), the solar corona density is 6.7×10^6 m^{-3}. The propagation path between Earth and Mars may closely approach the Sun when Mars is on the opposite side of the Sun relative to Earth. When the closest approach to the Sun is less than 4 R_s (that is, the propagation path passes through the deep solar corona close to superior conjunction), the solar wind plasma effect cannot be neglected. However, when the closest approach is greater than 4 R_s, because the solar wind plasma along the path is far less than the peak densities of both Martian and Earth ionospheres, the solar wind plasma effect is negligible.

References

Flock, W.L., Telecommunications in cometary environments, JPL Publication 81-84, Jet Propulsion Laboratory, Pasadena, CA, 1981.

Goldhirsh, J., and W.J. Vogel, Handbook of Propagation Effects for Vehicular and Personal Mobile Satellite Systems: Overview of Experimental and Modeling Results, APL-A2A-98-U-0-021, Applied Physics Laboratory, Laurel, MD, 1998.

Halliday, I, and B.A. McIntosh, Solid particles in the solar system, Symposium No. 90, Organized by the IAU in cooperation with COSPAR held at Ottawa, Canada, Aug. 27–30, 1979.

Smith, E.K., and R.E. Edelson, Radio propagation through solar and other extraterrestrial ionized media, JPL Publication 79-117, Jet Propulsion Laboratory, Pasadena, CA, 1980.

8. Summary and Conclusions

8.1 Ionospheric Effects

Because Mars has almost no intrinsic magnetic field, the solar wind can directly interact with the upper atmosphere. The height of the ionopause is controlled by the solar wind pressure. The dayside Martian ionosphere may be described using a simple Chapman layer model. The Martian dayside ionosphere has stable peak height and peak density. The peak height is between 120 and 130 km. On average, the Martian ionospheric plasma density is one order lower than Earth's and its TEC value is 50 times lower than Earth's ionospheric TEC. The Martian ionosphere is almost transparent to radio waves with frequencies above 450 MHz. For frequencies below 450 MHz, there is a progressive degradation of signal until the 4.5-MHz cut-off frequency where waves cannot pass through the ionosphere. Because the Martian ionosphere is similar to that of Earth in some aspects, we have used 0.5 dB for VHF-band signal and smaller losses for higher frequency bands. There is a comet-like structure to the nightside ionosphere extending several thousand kilometers in the antisolar direction. The plasma density in the nightside ionosphere is very low ($\leq 5 \times 10^3$ cm^{-3}). The nightside ionospheric profile often shows no dominant density peak and has large variations.

Recommendation: The Martian ionosphere may be used as a reflector for global communication. This is crucial for future Mars ground-to-ground communication. The Martian dayside ionosphere has a critical frequency of ~4.0 MHz for vertical incidence. This frequency is high enough to carry information. The stable condition in the dayside ionosphere also favors oblique incidence communication. Reflection off the Mars ionosphere can also provide trans-horizon (or beyond line of sight) communication for future Martian colonies, rovers, vehicles, and robots released from Mars landers. However, because of low usable frequency and very unstable condition, the nightside ionosphere has serious limitations for global communication.

Suggestions: We do not yet have any nightside ionosphere model. This is mainly because very few nightside ionospheric measurements are available. Thus, we suggest a detailed nightside ionosphere study. The MGS spacecraft is performing dayside ionospheric occultation and in-situ measurements. It will take some time to shift to the nightside. Previous occultation measurements showed that often no density peak was seen in the nightside ionosphere. We propose to drop a digital ionosonde to the Mars surface at a low latitude. The ionosonde will transfer daily ionospheric sounding data either to an orbiter or directly to Earth. We can have daily ionospheric data on peak height, critical frequency, etc. Through this study, we can assess stability of the nightside ionosphere for use as a radio wave reflector. Eventually, a model for the nightside ionosphere will be developed.

We still need to discover many things about the overall for the entire Martian ionosphere. To calculate low-frequency radio wave attenuation, we need to measure the ionospheric collision frequency. Ionospheric scintillation will be related to plasma irregularities and turbulence. For Earth during an ionospheric storm, absorption of VHF waves is enhanced. We are not sure that the same thing can happen in the Mars ionosphere. The ionospheric refractive index fluctuation and gradients will cause radio wave ray bending. We need to perform an accurate calculation of these effects.

8.2   Tropospheric Effects

The Martian troposphere is so thin that we expect it to have very little effect on radio wave propagation. The refractive index of the Martian troposphere at the surface is about two orders of magnitude smaller than that of Earth. Attenuation due to clouds and fog depends largely on their water content. So far we have little knowledge about such attenuation because these measurements are not yet available. Martian clouds are expected to have relatively less water liquid content than terrestrial clouds because measurements show that the clouds have low optical depth. At most, the Martian clouds are similar to terrestrial high-level cirrus clouds. Martian fog and aerosols (haze) also have a small optical depth. The total attenuation due to Martian clouds and fog should be less than 0.3 dB at Ka-band. There is a plasmasheath effect on communication during Martian atmospheric entry phase due to spacecraft impact ionization.

Recommendation: Even though the Martian tropospheric radio refractivity has a small value, it can still cause ray bending and multipath effects. We recommend performing an accurate calculation on excess phase and group delays (range and time delays). Other effects (such as range rate errors, appearance angle deviation, and defocusing loss on Mars) should also be estimated. Ice depolarization effects due to Martian clouds on radio waves are still unknown, but they are expected to be small because of the lower optical depth and the thinner cloud layer. Thus, the Martian atmospheric environment is also good for optical communications, except during dust storms. Even though Mars aerosols can cause some attenuation to a laser beam, this effect is very small, as compared with Earth.

For the communication blackout during Martian atmospheric entry phase, three solutions are recommended:

1) Place the antenna where the plasma is diluted, e.g., on the lee side; or communicate by a relay orbiter.

2) Inject some electron-absorbing liquid chemicals into the flowfield to neutralize the plasma.

3) Increase the frequency of transmission signals from X-band to Ka-band.

Future Martian atmospheric entry and manned landing programs should experimentally test to decide the best of these three options.

Suggestions: In the future, the amount of water or condensed CO_2 in the clouds should be measured accurately. Also, dust amounts in the atmosphere should be monitored, and the height in the atmosphere to which the dust extends should be determined. Absorption by water vapor, and the size and shape of the dust particles, should be measured. To study tropospheric scattering and scintillation, the tropospheric turbulence needs to be measured. We also need to estimate the effects of the plasma sheath due to the impact ionization on radio wave propagation during atmospheric entry. As on Earth, the entry blackout can temporarily shut down communications between the spacecraft and ground.

8.3   Gaseous Attenuation

Mars has an atmospheric gaseous attenuation of less than 1 dB in the microwave band, because the Martian atmosphere has very low concentrations of gaseous H_2O and O_2. Martian gaseous absorption is at least three orders of magnitude lower than at Earth. An accurate calculation for zenith opacity requires information about scale heights of H_2O and O_2. The ratio of total zenith absorption in the Earth atmosphere relative to Mars should be equal to the ratio of column number densities of H_2O and O_2 of Earth relative to Mars. We also do not know how high the gaseous attenuation can be in the infrared and optical frequency ranges because there are so many complicated absorption spectral lines in these frequencies.

Because the Mars troposphere consists of almost entirely dry air and the surface atmospheric water content is 3000 times lower than at Earth, the water absorption peaks in the spectrum are very low. Thus, the windows that on Earth are bounded by water lines become much wider. From 60 GHz to 300 GHz there is almost no attenuation. This feature is obviously in contrast to the Earth's situation, in which heavy rain and water vapor dominate the attenuation. The Martian atmosphere is dominated by CO_2 and N_2 gases. Under normal conditions, these gases do not have electric or magnetic dipoles, so they do not absorb electromagnetic energy. However, they may generate dipoles through collision and interaction with waves under a high-density condition. We often see that both gases have many absorption lines in the infrared and visible bands in the Earth atmosphere. It will be a research topic whether CO_2 and N_2 gases at the Martian surface can generate such dipoles. In our gaseous attenuation calculation, we have used an average surface value (300 ppm) for Martian water vapor, instead of a maximum value (400 ppm), which corresponds to the worst case. Actually, the exact amount of gaseous H_2O is still debatable. An accurate water vapor altitude profile at Mars is not available yet. A conservative estimate for the worst situation of Martian atmospheric attenuation is an increase by a factor 1.5.

8.4   Dust Storm Effects

Dust storms are the dominant factor in radio wave attenuation at Mars. A large dust storm can cause at least a 3-dB loss to Ka-band waves. For a normal dust storm, the attenuation is about 1 dB. The attenuation depends largely on dust mass loading, dust size distribution, etc. Currently we still have little information about these factors. Most large dust storms occur in the southern hemisphere during later spring and early summer when the southern hemisphere becomes suddenly hot.

Suggestions: In future missions, a number of dust storm parameters need to be further measured: Occurrence frequency, size, altitude, dust mass loading, dust size distribution, etc. Using this information, we will be able to accurately estimate dust storm effects on the wave attenuation. When the spacecraft lands in the southern hemisphere, at least a 3-dB margin should be considered for lander and rover communication.

8.5   Surface Geomorphologic Structures

Low-elevation-angle multipath fading due to surface rocks and terrain is another important impairment factor in wave propagation. Mars has a very complicated surface geomorphologic structure. When terrain or rocks block direct radio wave rays, a lander can sometimes still keep communication with a satellite through a diffracted or reflected ray. Diffracted wave signals will reduce intensities by at least a factor of 5 relative to direct signals. Reflected signal intensity strongly depends on the reflectivity of materials. Multipaths can also cause signal amplitude fading and attenuation due to phase shifts. Some studies of multipath effects for terrestrial canyon and hilly environments have been done. For an 870-MHz wave, attenuation is in a range of 2–7 dB, while for L-band (1.7 GHz), the attenuation is 2–8 dB. At higher frequencies, higher losses should be expected. Thus, surface rock attenuation will be potentially a significant attenuation source. We need to perform a similar multipath study on the Mars.

Suggestions: Low-elevation-angle fading is a potential communication problem for future Mars colonies and land vehicles. We need to study the fading due to rocks and terrain and to find their distributions in various scales. Currently we have very limited information regarding the surface and rock properties. We recommend that such types of measurements and experiments be included in future Mars missions.

8.6   Links between Mars and Earth

The minimum and maximum distances between Mars and Earth are, respectively, 55×10^6 and 400×10^6 km. Free space losses are 277 and 294 dB corresponding to these distances. In addition to free space loss, radio wave signals propagating from Mars to Earth suffer additional atmospheric losses at both planets. These combined atmospheric losses are about 8 dB under normal conditions. At Earth, for 99% of the time, weather conditions are such that total tropospheric attenuation for Ka-band is about 5 dB for a vertical propagation, while the comparable attenuation at Mars is about 3 dB. We have ignored medium loss in the interplanetary space because its effects are so small.

Finally, based on the Martian atmospheric environment, we strongly recommend using optical links for future Mars communications. Because of the thinner Martian atmosphere and the almost transparent Martian clouds, optical communication is almost perfect for links between Mars orbiters, between orbiters and landers, and even between Mars surface robots. Laser beams in the Martian atmosphere will have much less attenuation relative to those used in the Earth environment. We also suggest using low frequency (4.0 MHz) radio waves for Martian surface communication because the Martian ionosphere can effectively reflect these waves forward to areas beyond the line of sight. This will make Martian surface global communication possible.

8.7   Recommendations for Telecommunication Systems Engineer

8.7.1 Martian Ionospheric Effects

The Martian ionosphere only effects low-frequency waves (less than 450 MHz) and is almost transparent for high-frequency bands (S, X, and Ka). It has a loss of ~ 0.5 dB for the VHF (including UHF) band and negligible losses for higher frequency bands. The ionosphere has a critical frequency of ~4 MHz for vertical incidence. A wave with a 90-degree incidence angle and with frequency higher than the critical frequency will pass through the ionosphere unattenuated. The ionosphere can be used for future Mars surface trans-horizon communication. Martian ionospheric effects on one-way radio waves are summarized in Table 2-3 and plotted in Figure 8-1.

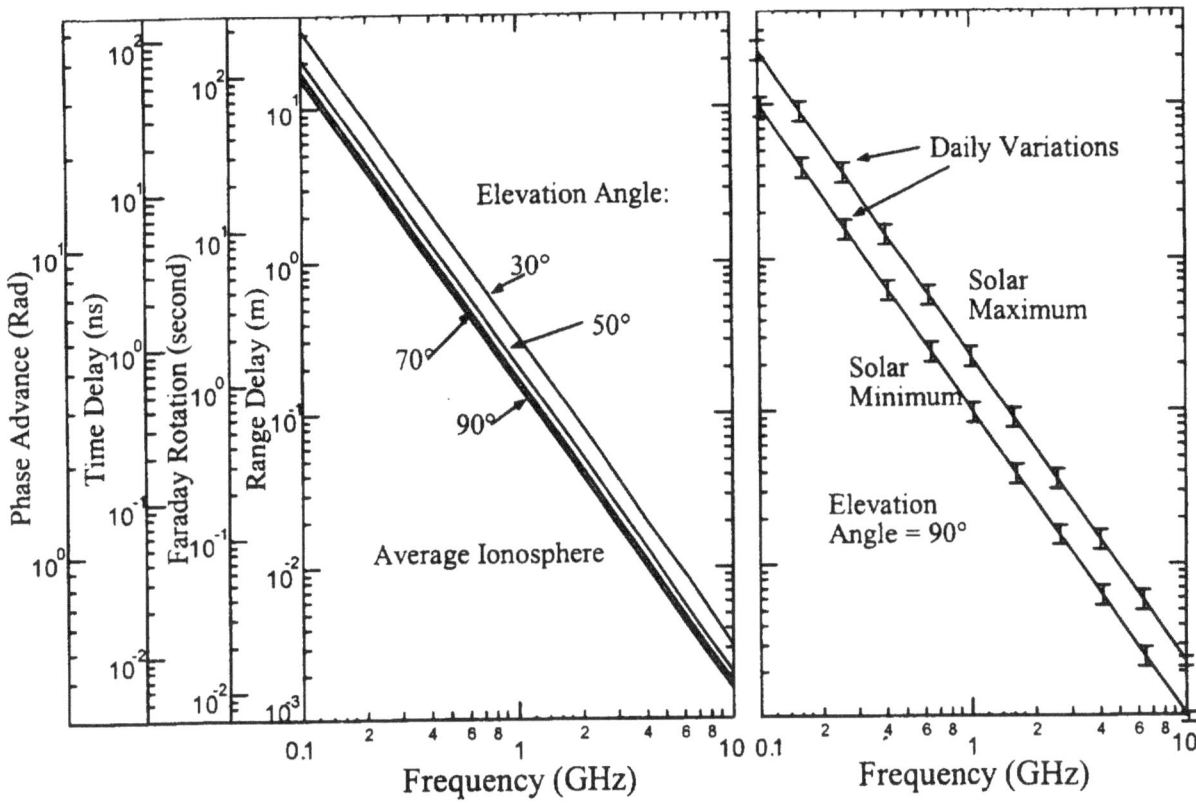

Figure 8-1. Martian Ionospheric Effects on One-Way Radio Wave Propagation of Various Frequencies. These effects include range delay, Faraday rotation, time delay, and phase advance. The left panel shows elevation angle effects for propagation through an average Martian ionosphere (between the solar maximum and minimum). The right panel shows solar cycle effects on the propagation with a zenith path. Daily variations (about ±15%) are also shown.

8.7.2 Martian Atmospheric Effects

The Martian atmosphere (or troposphere) is very thin and can be expected to have very little effect on radio wave propagation. Because Mars has a very low atmospheric pressure (less than 1% of Earth's), the Martian atmospheric radio refractivity is about two orders of magnitude smaller than that of Earth. Lower frequencies (UHF band) are expected to have very little refractive and scattering effects in the Martian troposphere. A high-frequency wave (above 1 GHz) may be bent or trapped by the vertical refractivity gradient when the wave incident angle is very close to the horizon.

8.7.3 Martian Cloud Effects on Wave Propagation

Optical depth is a measure of propagation loss. A transparent object has a small optical depth; an opaque object has a large optical depth. The optical depths of Martian clouds and fogs are about 1.0 at visual wavelengths. Thus, it is expected that they have little attenuation for microwave propagation. In the limiting case, the Martian clouds are expected to be similar to terrestrial high-level cirrus clouds. Martian aerosols (haze) have also been found to have a small optical depth (less than 0.5). The total attenuation due to Martian clouds, fog, and aerosols should be less than

0.3 dB at Ka-band. For lower frequency bands, the attenuation is almost negligible. All information about cloud attenuation is summarized in Tables 3-2, 3-3, and 3-4.

8.7.4 Martian Atmospheric Gaseous Attenuation

The atmospheric gaseous attenuation at Mars is greater at higher frequencies than at lower frequencies. However, the worst-case loss (at Ka-band) is still less than 1 dB. This is because the Martian atmosphere has very low concentrations of gaseous H_2O and O_2. Martian gaseous absorption is at least three orders of magnitude lower than that at Earth. Even though in the Martian atmosphere there is very little water vapor, it still dominates the entire gaseous attenuation, because compared to the Earth Mars has a much lower ratio of oxygen than water vapor. An accurate water vapor altitude profile at Mars is not yet available. Figure 8-2 shows gaseous attenuation for one-way radio wave path with two different elevation angles. We have used a 10-km scale height and a 300-ppm water vapor density, which may range from 100 to 400 ppm and represents an upper limit.

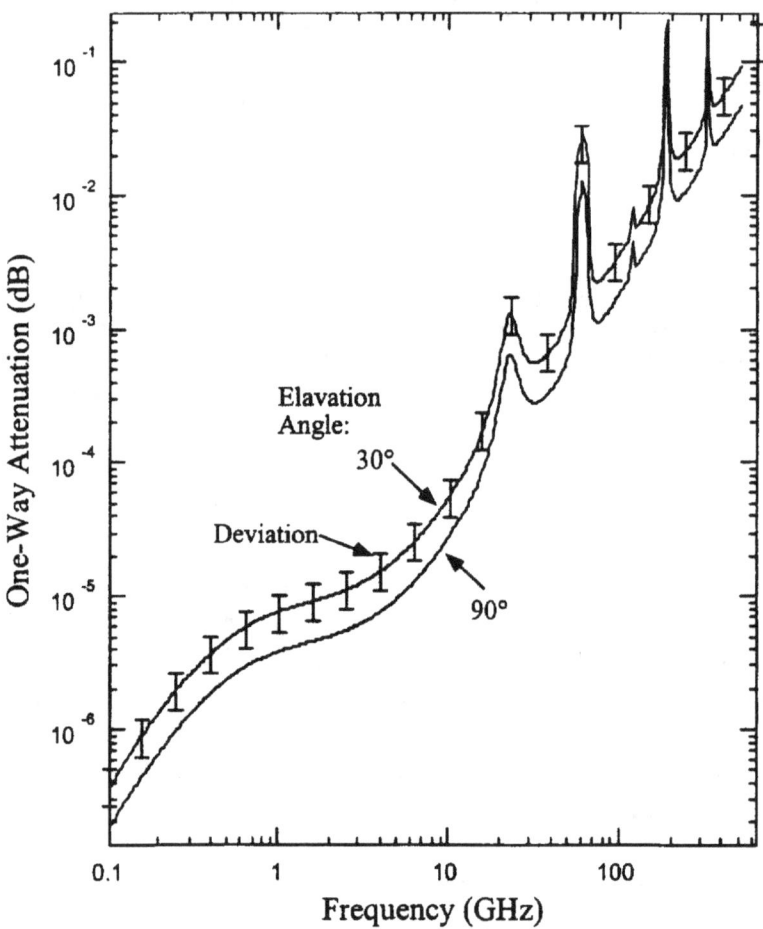

Figure 8-2. Martian Gaseous Attenuation for a One-Way Radio Wave Path through the Atmosphere for Two Different Elevation Angles (30° and 90°). The total attenuation showed here is mainly due to water vapor, even though the oxygen effect is also included. To calculate the attenuation, a 10-km scale height and a 300-ppm water vapor density have been used. The attenuation deviations for possible water vapor variations (about ±30%) are also shown.

8.7.5 Martian Dust Storm Effects

Dust storms on Mars can significantly affect a communication link. A large dust storm can cause at least a 3-dB loss at Ka-band. Lower frequency bands (UHF, S, and X bands) suffer less dust storm attenuation, which has a linear relationship with frequency and depends on the cube of particle size. Figure 8-3 shows Martian dust attenuation for a one-way radio wave path through a dust cloud with 10-km scale height for various elevation angles and particle sizes. The dust cloud parameters used for the calculation are summarized in Table 5-3. Most large storms occur in the southern hemisphere during later spring and early summer.

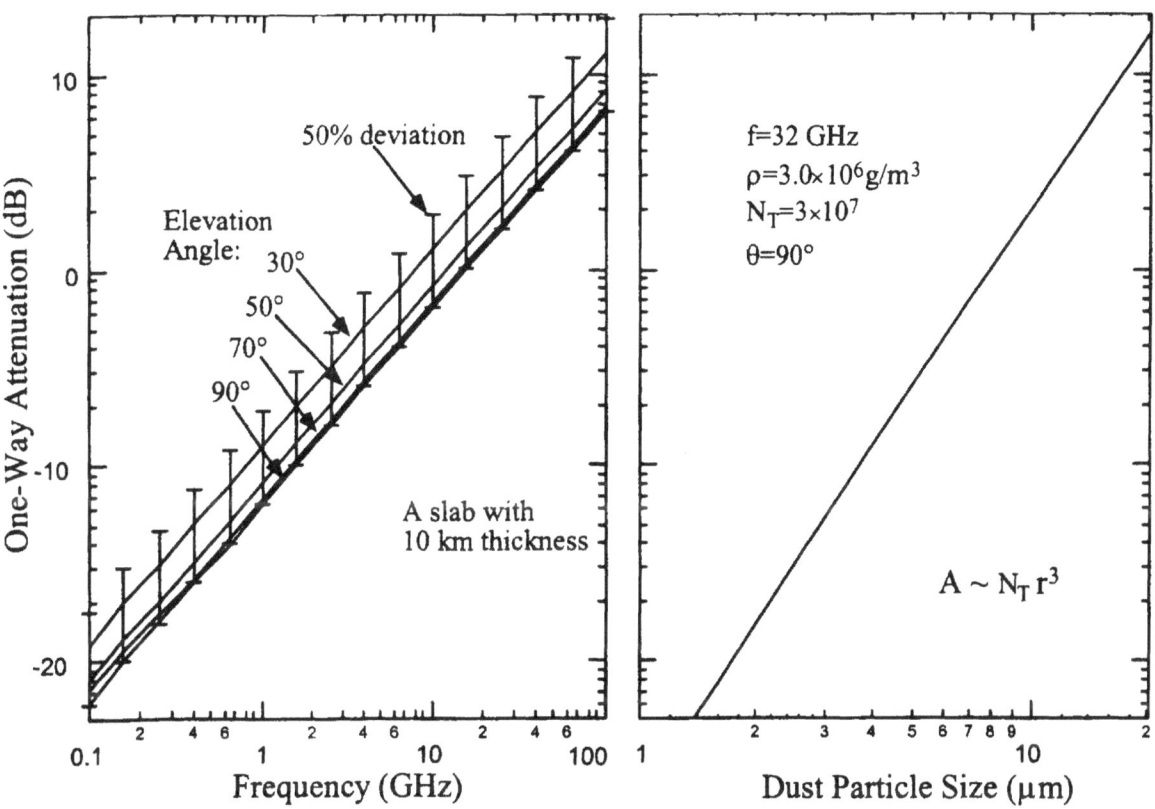

Figure 8-3. Martian Dust Attenuation for One-Way Radio Wave Path through a Dust Cloud for Various Elevation Angles and Dust Particle Sizes. To calculate the attenuation in left panel, a 10-km scale height and a 10-μm particle radius have been used. The attenuation deviations due to possible particle size variations (about ±50%) are also shown. Right panel shows the attenuation dependence on dust particle size for a 32-GHz radio wave signal with a zenith path.

8.7.6 Communication Blackout during the Martian Atmospheric Entry Phase

When a high-speed spacecraft enters the Martian atmosphere, a plasma sheath is formed in the front of the spacecraft due to the impacting ionization. This can cause a communication blackout, the extent of which depends on the communication frequency. A 30-s communication disruption was observed with the Mars Pathfinder (X-band), and a 1-minute blackout was experienced by both Viking Landers (UHF). If the frequency of a communications signal is higher than the critical frequency of the surrounding plasma, a radiowave can pass through the plasma sheath freely, and there will be no communication disruption. It is believed that this is the case at Ka-band.

9. Acronyms

ASI	Atmospheric Structure Instrument (on Viking Lander and Mars Pathfinder)
AU	astronomical unit, average distance between Sun and Earth (1.46×10^8 km)
b	bar = $1 \times 10^5 \, Pa$
BS	bow shock
CARV	cruise and relay vehicle
CCIR	International Radio Consultative Committee
CNES	French Space Agency
COSPAR	Committee on Space Research
EUV	extreme ultraviolet
ESA	European Space Agency
DOY	day of year
JPL	Jet Propulsion Laboratory
HDO	hydrogen-deuterium-oxygen
HST	Hubble Space Telescope
IMP	Imager for Mars Pathfinder
IMF	interplanetary magnetic field
ISM	Infrared Spectrometer (on Phobos 2)
IRTM	(Viking) Infrared Thermal Mapper experiment
MAGE	Mars Airoborne Geophysical Explorer
MET	Meteorology Package (on Mars Pathfinder)
MGS	Mars Global Surveyor
MPB	magnetic pile-up boundary
MPF	Mars Pathfinder
MOC	Mars Orbiter Camera
MOLA	Mars Orbiter Laser Altimeter (on MGS)
NASA	National Aeronautic and Space Administration
nT	nanotesla
ppm	parts per million
NOAA	National Ocean and Atmosphere Administration
rms	root of mean square
	R_s solar radius (6.96×10^5 km)

S.T.P.	standard temperature and pressure
SZA	solar zenith angle
TEC	total electron content
UAMS	Upper Atmosphere Mass Spectrometer (on Viking)
UHF	ultra-high frequency
UTC	Universal Time Coordinates (Greenwich time)
VHF	very high frequency
VL	Viking Lander
WFPC	Wide Field Planetary Camera (on Hubble Space Telescope)

10. Index

albedo: ratio of the flux reflected by the planet to the total incident flux, see pp72 and 82

aerosol: see pp27 and 82

aphelion: The farthest point of the Mars orbit from the Sun, see p2

aerobraking: Aerobraking is a technique to reduce the amount of fuel required to slow down the Mars Global Surveyor spacecraft as it approaches Mars. Just like an airplane uses spoilers and flaps to slow down prior to landing, the MGS spacecraft used the drag of the Martian atmosphere on its solar panels to slow down as an alternative to using thrusters which would have required extra fuel and, therefore, extra weight and cost. The duration of the aerobraking phase is directly related to how fast Mars' relatively thin atmosphere reduces the spacecraft's velocity, see p10.

bow shock: a thin current layer across which the flow (solar wind) velocity drops, and the plasma is heated, compressed, and deflected around the obstacle. Ahead of the bow shock the velocity component along the shock normal is supersonic and behind it, subsonic, see pp7 and 38.

boundary layer: layer of fluid flowing near a surface, which, because of the effects of viscosity, flows at reduced speed compared to the relativity inviscid flow outside the layer, see p38.

Chapman layer: An ionospheric layer which can be described by Chapman theory, in this layer, plasma (ion) density peaks at an altitude where solar extreme ultraviolet ionization rate and the recombination rate reach a balance, see pp9 and 10.

chemical equilibrium: see p39

cirrus cloud: see p34

communication blackout: a temporarily disruption of communication. When a high speed (supersonic) spacecraft enters the Martian atmosphere, because its flight speed is much greater than the local speed of sound, a plasmasheath forms around the spacecraft resulting from thermal ionization of the constituents of the air as it is compressed and heated by the strong bow shock or heated within the boundary layer next to the surface. Because the plasma density surrounding the spacecraft is very high, the communications are disrupted during the entry phase. This phenomena is usually known as a blackout, see p38.

critical frequency: also called as the local plasma frequency (f_p), where f_p (MHz) = 9.0×10^{-3} $N^{1/2}$ (cm^{-3}). For $f < f_p$ a plasma behaves like a conductor (radio wave will be reflected or absorbed by plasma), while for $f > f_p$ the plasma is practically transparent, see pp15, 16, and 38.

dielectric permitivity: see p32

diffraction: see pp72 and 78

dust devil: swirling, vertical updraft of air developed by local heating of the air above the flat desert floor. Dust devils are spinning columns of air that move across the landscape and look somewhat like miniature tornadoes. Dust devils are a common occurrence in dry and desert landscapes on Earth as well as Mars. They form when the ground heats up during the day, warming the air immediately above the surface. As the warmed air nearest the surface begins to rise, it spins. The spinning column begins to move across the surface and picks up loose dust. The dust makes the vortex visible and gives it the "dust devil" or tornado-like appearance. On Earth, dust devils typically last for only a few minutes, and the same is probably true for Mars. see pp27 and 36.

dust storms: atmospheric dust hazes or clouds, consisting of materials of silt (1/16 – 1/256 mm) and clay (< 1/256 mm) size blown by the wind, see pp59-70.

eccentricity of Mars orbit: the amount by which the orbit deviates from circularity: $e = c/a$, where c is the distance from the center to a focus and a is the semimajor axis; or $e = 1 - q/a$, where q is the periapsis distance, see p2.

hop distance: a radio wave obliquely launched from the ground will be reflected by the ionosphere, when its frequency is less than the ionospheric critical frequency. The wave will leap a hop distance after it reaches the ground again by $l = 2\,h\,tan\,\theta_0$ (where θ_0 is wave launch angle and h is ionospheric height), see p16.

Huygen's principle: Every elementary area of a wavefront can be regarded as a source of secondary spherical waves, see p78.

interplanetary dust: see p90

ionosphere: a layer consisting of ionized gases (plasma) created by solar extreme ultraviolet flux, which usually extends from 100 km to several hundred kilometers in altitude, see pp7-20

ionopause: the boundary between the solar wind plasma and the Martian ionospheric plasma, see pp8 and 12

ionosheath (magnetosheath or planetsheath): the region between a planetary bow shock and ionopause in which the shocked solar wind plasma flow around the ionosphere, see pp8 and 12

ionotail (magnetotail): a comet-like region of nightside ionospheric plasma in the wake of a planetary obstacle to a plasma flow, see p8

Kelvin temperature: T K= T°C + 273.15°.

Ls: Ls is the aerocentric longitude of the Sun as measured in a Mars-centered fixed coordinate system, often used as an angular measure of the Mars year (Ls = 90°, 180°, and 270°

corresponding to the beginning of southern winter, spring, and summer, respectively), see p60.

magnetic barrier: see p8

multipath: see pp28 and 72

Neper: a dimensionless unit for optical depth. It can be expressed in (logarithms to base e), A $(Np) = 4.34$ A (dB), see p47

nominal mean model: see p22

optical depth: Optical depth, τ, is a measure of attenuation over the entire path taken from the ground to space. Optical depth may be obtained through the following measurements. The power received, P_r, is the power transmitted, P_t, multiplied by the attenuation: $P_r = P_t\, e^{-\tau}$ (i.e., $\tau = ln(P_t\,/\,P_r)$). Thus, a transparent object has small optical depth, while an opaque object has large optical depth, The term "opacity" is commonly also used, see pp34, 35, and 68.

opacity: a measure of the ability of an atmosphere to absorb or scatter radiation, see pp68 and 82

opposition: When Earth passes between Mars and the Sun, there is a minimum distance between the two planets, see pp 1 and 87.

perihelion: The closest point of Mars orbit from the Sun, see p 2

plasmasheath: similar to the ionosheath, a region between a bow shock and boundary layer in which the shocked and heated plasma flow around the obstacle, see p 38.

polar cap: see pp 74 and 75

polar hoods: see p80

radiation inversion: see p27

Rayleigh distribution: p85

Rayleigh scattering: p32

reflection: see pp72 and 78

reflectivity: the reflectivity (R) is the ratio of returned laser energy to the emitted laser energy. R is affected by the surface albedo (A) of the underlying terrain and extinction of the photons from the laser beam by atmospheric aerosols, see p82.

refractivity: tropospheric radio refractivity is defined as the difference between the gas refractive index and unity, that is, $N = (n-1) \times 10^6$ (N unit), see p 5.

residual caps: Residual caps are inner parts of polar ice caps which do not change with seasons, that is, they are the smallest polar ice cap. The southern residual cap is about 350 km across, compared with 1,000 km for the remnant northern cap. The northern residual cap is almost certainly water ice, while the southern cap probably consists predominantly of CO_2 ices, see pp 75 and 76.

Ricean distribution: p 85

scintillation: see p 30

seasonal caps: Seasonal caps are the Mars polar ice caps, which advance and recede with seasons. Martian seasonal polar ice caps have their maximum sizes about 2000 km across during winter, see pp 75, 76, and 77.

sol: one Martian solar day, equivalent to 24.66 terrestrial hours, see p 27

solar corona: see pp1 and 90 and 91

solar wind: see pp 7 and 90

stagnation region: see pp 39 and 40

superior conjunction: The two planets are on opposite sides of the Sun. There is a maximum distance between two planets, see pp 1 and 87.

surface (6.1 mb level): the 0-km altitude, which is defined as the Mars reference surface at atmospheric pressure 6.1 mb level (610 Pa), see pp 25 and 74

terminator: see p 12

troposphere: see pp 22–30

Valles Marineris: On the east side of Tharsis and just south of the equator, between longitudes 30°W and 110°W, there are several enormous, interconnected canyons, which have been collectively called Valles Marineris. The Valles Marineris is the most spectacular geologic feature on Mars. The canyon is 4000 km long, 150 km wide, and 10 km deep, see pp74 and 81.

wake region: see pp39 and 40